Friederike Brendel

Millimeter-Wave Radio-over-Fiber Links based on Mode-Locked Laser Diodes

Karlsruher Forschungsberichte
aus dem Institut für Hochfrequenztechnik und Elektronik

Herausgeber: Prof. Dr.-Ing. Thomas Zwick

Band 68

Millimeter-Wave Radio-over-Fiber Links based on Mode-Locked Laser Diodes

by
Friederike Brendel

Dissertation, Karlsruher Institut für Technologie (KIT)
Fakultät für Elektrotechnik und Informationstechnik, 2013

Impressum

Karlsruher Institut für Technologie (KIT)
KIT Scientific Publishing
Straße am Forum 2
D-76131 Karlsruhe
www.ksp.kit.edu

KIT – Universität des Landes Baden-Württemberg und
nationales Forschungszentrum in der Helmholtz-Gemeinschaft

KIT Scientific Publishing 2013
Print on Demand

ISSN 1868-4696
ISBN 978-3-86644-986-2

Editor's Preface

We are being exposed to a revolution which is taking us from a world where we used to be constrained to communicate over fixed telephone lines to one where a mobile communication environment has become a reality for voice and data. Recently, voice services have even taken a back seat in favor of wireless data services, which have come to dominate all mobile communication networks. Besides cellular data services, local area networks play a key role in providing both commercial and private users with wireless access at ever increasing data rates, in particularly indoors. Conventional wireless local area network services operating in the lower GHz frequency range where the electromagnetic spectrum is densely occupied currently offer data rates up to 600Mbit/s under optimal conditions. In order to further increase the data rate, carrier frequencies in the millimeter-wave range have to be used since only there the necessary bandwidth is available.

In Europe, up to 9 GHz bandwidth are available for unlicensed short-range wireless data communication in the 60 GHz region. However, the relatively high frequency range leads to reduced radio coverage due to an increased path loss and also higher wall penetration losses. As a direct consequence, each room is required to be equipped with its own 60 GHz access point, which, in turn, facilitates the re-allocation of frequencies and thus enables the creation of a dense user environment using simple access schemes. In order to be cost-effective, these access points should be of low complexity and easy to install. Signal distribution between the central station and the access points can be realized using optical fiber networks which are flexible and exhibit low loss as well as high bandwidth characteristics.

The lowest complexity in the access point can be achieved when the analog 60 GHz radio signal is already generated in the central station and delivered via the optical fiber to the access point. Such a radio-over-fiber link was addressed in the dissertation of Ms. Friederike Brendel. A particular challenge is the generation and modulation of the analog 60 GHz signal by means of lightwave signals. In her dissertation Ms. Brendel investigated techniques relying on signal generation by passively mode locked laser diodes in millimeter-wave radio-over-fiber systems. The core of the work is a full system-level study on different link architectures combining the three necessary technical fields of photonics, microwaves and communications.

A clear connection could be made between the physical behavior of the photonic devices and the overall link performance, which is greatly impressive given the complexity of the opto-electronic system. An aspect which should receive particular attention is that of the typically limited stability of the signal generated by such mode-locked laser devices, making their application in commercial systems seem premature. Ms. Brendel's invention of a novel method for frequency stabilization of the resulting millimeter-wave signal therefore represents a major scientific contribution which makes Ms. Brendel's work very valuable both component designers and system engineers. It represents an essential enhancement of the state of the Art, and I am positive that her work will attract much interest from the research community worldwide. My personal wish for Ms Brendel is that her creativity and great innovative capacity will continue to earn her both scientific and economic success.

Prof. Dr.-Ing. Thomas Zwick
– Director of the IHE –

Forschungsberichte aus dem
Institut für Höchstfrequenztechnik und Elektronik (IHE)
der Universität Karlsruhe (TH) (ISSN 0942-2935)

Herausgeber: Prof. Dr.-Ing. Dr. h.c. Dr.-Ing. E.h. mult. Werner Wiesbeck

Band 1   Daniel Kähny
Modellierung und meßtechnische Verifikation polarimetrischer,
mono- und bistatischer Radarsignaturen und deren Klassifizierung
(1992)

Band 2   Eberhardt Heidrich
Theoretische und experimentelle Charakterisierung der
polarimetrischen Strahlungs- und Streueigenschaften von Antennen
(1992)

Band 3   Thomas Kürner
Charakterisierung digitaler Funksysteme mit einem breitbandigen
Wellenausbreitungsmodell (1993)

Band 4   Jürgen Kehrbeck
Mikrowellen-Doppler-Sensor zur Geschwindigkeits- und
Wegmessung - System-Modellierung und Verifikation (1993)

Band 5   Christian Bornkessel
Analyse und Optimierung der elektrodynamischen Eigenschaften
von EMV-Absorberkammern durch numerische Feldberechnung (1994)

Band 6   Rainer Speck
Hochempfindliche Impedanzmessungen an
Supraleiter / Festelektrolyt-Kontakten (1994)

Band 7   Edward Pillai
Derivation of Equivalent Circuits for Multilayer PCB and
Chip Package Discontinuities Using Full Wave Models (1995)

Band 8   Dieter J. Cichon
Strahlenoptische Modellierung der Wellenausbreitung in
urbanen Mikro- und Pikofunkzellen (1994)

Band 9   Gerd Gottwald
Numerische Analyse konformer Streifenleitungsantennen in
mehrlagigen Zylindern mittels der Spektralbereichsmethode (1995)

Band 10   Norbert Geng
Modellierung der Ausbreitung elektromagnetischer Wellen in
Funksystemen durch Lösung der parabolischen Approximation
der Helmholtz-Gleichung (1996)

Band 11   Torsten C. Becker
Verfahren und Kriterien zur Planung von Gleichwellennetzen für
den Digitalen Hörrundfunk DAB (Digital Audio Broadcasting) (1996)

Forschungsberichte aus dem
Institut für Höchstfrequenztechnik und Elektronik (IHE)
der Universität Karlsruhe (TH) (ISSN 0942-2935)

Forschungsberichte aus dem
Institut für Höchstfrequenztechnik und Elektronik (IHE)
der Universität Karlsruhe (TH) (ISSN 0942-2935)

Forschungsberichte aus dem
Institut für Höchstfrequenztechnik und Elektronik (IHE)
der Universität Karlsruhe (TH) (ISSN 0942-2935)

Forschungsberichte aus dem
Institut für Höchstfrequenztechnik und Elektronik (IHE)
der Universität Karlsruhe (TH) (ISSN 0942-2935)

Fortführung als
"Karlsruher Forschungsberichte aus dem Institut für Hochfrequenztechnik
und Elektronik" bei KIT Scientific Publishing
(ISSN 1868-4696)

Karlsruher Forschungsberichte aus dem
Institut für Hochfrequenztechnik und Elektronik
(ISSN 1868-4696)

Herausgeber: Prof. Dr.-Ing. Thomas Zwick

Die Bände sind unter www.ksp.kit.edu als PDF frei verfügbar
oder als Druckausgabe bestellbar.

Karlsruher Forschungsberichte aus dem
Institut für Hochfrequenztechnik und Elektronik
(ISSN 1868-4696)

Millimeter-Wave Radio-over-Fiber Links based on Mode-Locked Laser Diodes

Zur Erlangung des akademischen Grades eines

DOKTOR-INGENIEURS

von der Fakultät für
Elektrotechnik und Informationstechnik
des Karlsruher Instituts für Technologie (KIT)

genehmigte

DISSERTATION

von

Dipl.-Ing. Friederike Cornelia Brendel

geb. in Crailsheim

Tag der mündlichen Prüfung: 23. 01. 2013

Hauptreferent: Prof. Dr.-Ing. Thomas Zwick
Korreferent: Prof. Dr. Béatrice Cabon

Submitted under a Co-Supervision Agreement
In Partial Fulfillment of the Requirements for the Jointly Awarded Degrees

DOCTEUR DE L'UNIVERSITÉ DE GRENOBLE

at Institut Polytechnique de Grenoble (Grenoble INP),
Grenoble, France

and

DOKTOR-INGENIEUR

at Karlsruher Institut für Technologie (KIT),
Karlsruhe, Germany

MEMBERS OF THE FRENCH EXAMINING BOARD

Prof. Dr. Catherine Algani	Présidente et Rapportrice
Prof. Dr.-Ing. Martin Vossiek	Rapporteur
Prof. Dr.-Ing. Sören Hohmann	Membre du Jury
Prof. Dr. Béatrice Cabon	Directrice de thèse
Prof. Dr.-Ing. Thomas Zwick	Directeur de thèse
Dr. Julien Poette	Co-encadrant, Invité du Jury

MEMBERS OF THE GERMAN EXAMINING BOARD

Prof. Dr.-Ing. Thomas Zwick	Hauptreferent
Prof. Dr. Béatrice Cabon	Korreferentin
Prof. Dr.-Ing. Ellen Ivers-Tiffée	Prüfungsvorsitz
Prof. Dr.-Ing. Sören Hohmann	Prüfer KIT
Prof. Dr.-Ing. Christian Koos	Prüfer KIT

Acknowledgements

This dissertation is the product of a particularly fruitful cooperation between two partners: The group of Prof. Dr. Béatrice Cabon at the *Institut de Microélectronique, Electromagnétisme et Photonique* (IMEP-LaHC) at Grenoble INP, in Grenoble, France, and the group of Prof. Dr.-Ing. Thomas Zwick at the *Institut für Hochfrequenztechnik und Elektronik* (IHE) at the Karlsruhe Institute of Technology in Karlsruhe, Germany. In French, such a project is referred to as "cotutelle de thèse"; however, an appropriate translation needs yet to be invented in German. Many different people have contributed to the success of this *cotutelle*:

First and foremost, I would like to thank my academic supervisors, Prof. Dr. Béatrice Cabon from the IMEP-LaHC and Prof. Dr.-Ing. Thomas Zwick, Head of the IHE, as well as Dr. Julien Poëtte from the IMEP-LaHC, for making this German-French cooperation a reality. Throughout the time of my research, they have supported me both on a professional and on a personal level, and I am well aware of the fact that such encouragement is never to be taken for granted.

I am very grateful to Prof. Dr. Catherine Algani of the Laboratoire ESYCOM (Electronique, Systèmes de Communication et Microsystèmes), Conservatoire National des Arts et Metiers (CNAM) in Paris, and Prof. Dr.-Ing. Martin Vossiek of the Friedrich-Alexander Universität Nürnberg-Erlangen who accepted to read and comment on the manuscript, and to travel to Karlsruhe to assist the defense as external reviewers. Prof. Dr. Algani amiably presided the French jury, while the German jury was headed by Prof Dr.-Ing. Ellen Ivers-Tiffée. Prof Dr.-Ing. Sören Hohmann kindly accepted the double charge of serving on both examination boards, and the German jury was completed by Prof. Dr.-Ing. Christian Koos. I would also like to thank Prof. Dr. Gérard Ghibaudo, Head of the IMEP-LaHC for welcoming me among the group of PhD students at IMEP-LaHC, as well as vice-director Prof. Dr. Jean-Emmanuel Broquin, for his tremendous support and valuable advice in times when it was most needed. I also much appreciated the conversations with Prof. Dr.-Ing. Ingmar Kallfass, another Francophile, from whose experience and advice I have been able to profit numerous times during my time as a doctoral student.

Furthermore, I would like to express my warmest gratitude to those who have made an experimenter's life easier, namely, the lab engineers and technicians whom I could always count on for both professional help and amicable discussion: Mr. Nicolas Corrao, Mr. Oliver Drouin and Mr. Grégory Grosa from the IMEP-LaHC and Mr. Ronald Vester from the IHE. I am grateful to Dave Mac-Farquhar who amiably proof-read the manuscript for language - so that I am now confident enough to publish my work in English.
My thesis project was funded by the French-German University (Deutsch - Französische Hochschule DFH, or Université Franco-Allemande UFA) at Saarbrücken, Germany. Their generous support made it possible to travel between Karlsruhe and Grenoble as well as to conference sites worldwide.

Throughout the past years, I was lucky to meet a number of colleagues who have become dear friends to me. From among the Grenoble crowd, my very special thanks goes to my office mates Marco Casale, Hana Ouslimani, Elsa Jardinier, François Parsy, as well as to Gabriel Rodriguez dos Campos, Flora Paresys, Vincent Dobremez, Léonce Mutwewingabo, and Vitor Freitas, as well as to Armin Schimpf, Thomas Nappez, Bertrand Charlet and Quentin Rafhay. Among the collegues and friends in Karlsruhe, I am particularly indebted to Steffen Scherr who proof-read the manuscript with impressive patience and accuracy. I would also like to thank my office mate Daniel Müller, as well as Jochen Antes and Ulrich Lewark - they will know what for, - and the entire IHE staff for their enthusiasm and team spirit. For all those hours spent at the desks, *en manip*, in the mountains around Grenoble, on ski and snowboard, in the forests in and around Karlsruhe, in the movie theaters - but above all, in your company, I thank you all. Finally, I would like to thank my closest friends and family, above all, Matthias Kühne, for bearing with me throughout this time.

Abstract

This dissertation is related to the search for an economically sustainable solution for high data rate (several Gbps to several tens of Gbps) personal area networks operating in the millimeter-wave region around 60 GHz. If such networks supply a large number of users, they need to encompass a multitude of antenna points in order to assure wireless access to the network. With the aim of reducing the cost of an antenna module, the networks should at best provide quasi "ready-to-radiate" signals to the modules, i.e. at millimeter-wave carrier frequencies. Thanks to their low transmission loss and their high bandwidth, optical fiber distribution architectures represent a promising solution. The technique is referred to as the so-called "radio-over-fiber" approach whereby the analog radio signal will be transported to the access point by an optical wave. The challenge hereby is the generation and modulation of an optical signal by a millimeter-wave radio signal using preferably cost-efficient system components. The technique proposed herein is based on the use of mode-locked laser diodes which can generate signals at very high frequencies under the condition of continuous current supply. Mode-locked laser diodes can be modulated both directly and externally. The diodes employed in this work are based on so-called *quantum dots* (or *quantum dashes*); these are material structures which are themselves still subject to intensive physical research. Signals at millimeter-wave frequencies (around 60 GHz) can easily be generated by such lasers. However, their frequency and phase stability is as yet limited. In the context of radio-over-fiber communication systems, these structures have not yet been studied in detail.

In the course of this dissertation, several aspects are considered. A first system study treats the basic properties of a system built from this type of laser source (available signal power, system noise figure, linearity etc.). A second study is devoted to an investigation of propagation effects like dispersion, which considerably influence the attainable transmission distances. An essential result of both studies is the importance of limiting the laser spectrum to a small number of laser modes for an optimization of link gain, generated RF power, and system noise figure. A third study deals with the limited frequency and phase stability which turn out to be critical factors for transmission quality. The study of several generations of quantum dot/dash lasers has revealed that the problems of frequency and phase noise persist and cannot be solved using classical techniques involving e.g. conventional phase-locked loops. In this dissertation, a solution is presented which not only allows a more precise adjustment of the laser frequency (precision in the order of Hz to kHz) than that given by the manufacturing process of the laser (precision in the order of GHz), but also enables a stabilization of frequency and phase.

Résumé

Ce travail de thèse s'inscrit dans la recherche des solutions économiquement viables pour des réseaux personnels à hauts débits (plusieurs Gbps à plusieurs dizaines de Gbps) opérationnels en bande millimétrique autour de 60 GHz. Au cas où ces réseaux servent un nombre élevé d'utilisateurs, ils comprendront une multitude d'antennes afin d'assurer l'accès sans fil rapide. Afin de réduire au maximum le coût d'un module d'antenne, les réseaux doivent fournir un signal analogue à des porteuses millimetriques. Une solution prometteuse pour les systèmes de distribution qui correspond à ces besoins sont des structures à fibre optique, laquelle permet une transmission à faibles pertes et à haute bande passante. On parle de l'approche "radio-sur-fibre" (en anglais, *radio-over-fiber*). La problématique est de pouvoir générer et moduler un signal aux fréquences millimétriques lors de la transmission optique - et ce avec des composant bas coûts. La technique utilisée dans le cadre de cette thèse est l'emploi des diodes laser à verrouillage de modes. Ces derniers vont pouvoir générer des hautes fréquences tout en ne nécessitant qu'une alimentation continue, et ils peuvent être modulés de manière directe ou externe. Les lasers à semi-conducteurs employés ici sont d'une génération encore à l'état d'étude puisqu'il s'agit des lasers à boites (ou îlots) quantiques. Ces lasers ont montrés de très bonnes capacités à générer des signaux électriques aux fréquences autour de 60 GHz, bien qu'ayant encore, pour l'instant, à une stabilité de fréquence (ou de phase) limitée. Dans le cadre des systèmes de communication opto/micro-ondes, peu de travaux approfondis ont été menés sur ces structures.

Au cours de cette thèse, plusieurs études ont été effectuées. La première porte sur les propriétés générales d'un système construit à partir de ce type de laser (puissances disponibles, figure de bruit, linéarité etc.). Une deuxième étude a été consacrée aux effets de la propagation des signaux dans les systèmes basés sur les lasers à verrouillage de modes, notamment de la dispersion chromatique laquelle a un effet considérable sur les distances de transmission. Les deux études mettent en avant l'importance d'une limitation du nombre de modes générés par la diode laser afin d'optimiser non seulement le gain du lien et la puissance RF récupérée, mais aussi la figure de bruit du système. Lors d'une troisième étude, la stabilité en fréquence/phase s'est révélée critique, car le bruit de fréquence/phase limite la qualité de la transmission en introduisant un plancher d'erreur même pour des rapports signal-a-bruit très élevés. Des différentes générations de lasers à boites (îlots) quantiques et à verrouillage de modes ont été testées. Le problème du bruit de fréquence et de phase persiste et ne peut pas être résolu en utilisant les techniques classiques comme les boucles à verrouillage de phase convention-nelles. Une solution pour ce problème a été développée pour les systèmes de transmission; elle permet simultanément un ajustement de fréquence supérieure (précision de quelques Hz à quelques kHz) à celle donnée par le processus de fab-rication des diodes lasers (précision de quelques GHz), ainsi qu'une stabilisation de fréquence et de phase.

Kurzzusammenfassung

Die vorliegende Dissertation steht im Zusammenhang mit der Suche nach wirtschaftlich tragfähigen Lösungen zum Aufbau hochdatenratiger Heimnetzwerke (einige Gbps bis einige zehn Gbps), so genannter *Personal area*-Netzwerke im Millimeterwellenbereich um 60 GHz. Sollen diese Netze eine große Anzahl von Nutzern versorgen, wird eine Vielzahl von Zugangspunkten - also Antennenmodulen - benötigt, um den drahtlosen Netzanschluss zu ermöglichen. Um die Kosten eines Antennenmoduls soweit wie möglich zu senken, sollen die Netze quasi "abstrahlfertige" Signale an die Module liefern, d. h. auf Trägerfrequenzen im Millimeterwellenbereich. Glasfaserbasierte Verteilsysteme werden dank ihrer geringen Leitungsverluste und ihrer hohen Bandbreite diesem Anspruch gerecht. Man spricht hier vom so genannten *Radio-over-fiber*-Ansatz, wobei das analoge Signal von einer optischen Welle zum Zugangspunkt transportiert wird. Die Herausforderung liegt hierbei in der Generierung und Modulation eines optischen Signals mit einem Nutzsignal im Millimeterwellenbereich - und das unter Verwendung möglichst kostengünstiger Komponenten. Die hier vorgeschlagene Technik basiert auf der Nutzung von modengekoppelten Laserdioden, welche allein bei Gleichstromversorgung Signale bei hohen Frequenzen erzeugen und sowohl direkt als auch extern moduliert werden können. Die Dioden, welche hier zur Verwendung kommen, basieren auf so genannten Quantenpunkten (englisch: *quantum dot/quantum dash*); es sind Strukturen, die selbst noch Gegenstand intensiver physikalischer Forschung sind. Signale bei Frequenzen um 60 GHz können leicht von diesen Lasern erzeugt werden, wenn auch bisher nur bei begrenzter Frequenz- und Phasenstabilität. Im Kontext von Radio-over-fiber-Systemen wurden diese Strukturen noch nicht untersucht.

Im Rahmen dieser Dissertation wurden mehrere Aspekte betrachtet. Eine erste Systemstudie behandelt die grundlegenden Eigenschaften eines Systems, welches auf dieser Art von Lasern basiert (verfügbare Leistung, Rauschzahl, Linearität usw.). Eine zweite Untersuchung ist der Erforschung von Ausbreitungseffekten wie etwa Dispersion gewidmet, welche die erreichbaren Entfernungen maßgeblich beeinflusst. Ein wesentliches Ergebnis beider Studien ist die Relevanz einer Begrenzung des Laserspektrums auf wenige Moden zur Optimierung von Gewinn, Hochfrequenz-Leistung und Rauschzahl. Eine dritte Studie untersucht die Frequenz-und die Phasenstabilität, welche sich als kritisch für die Übertragungsqualität erweisen. Die Untersuchung von mehreren Generationen von modengekoppelten Quantenpunktlasern hat ergeben, dass das Problem des Frequenz- und Phasenrauschens fortbesteht und nicht auf konventionellem Weg wie z.B. durch die Verwendung von klassischen Phasenregelkreisen gelöst werden kann. Im Rahmen der Arbeit wurde eine Lösung für dieses Problem gefunden, welche erstens eine bessere Feineinstellung der Frequenz erlaubt (Genauigkeit von Hz bis kHz), als sie durch den Laserfertigungsprozess gegeben ist (Genauigkeit von GHz), und zweitens eine Stabilisierung von Frequenz und Phase ermöglicht.

Contents

List of Abbreviations and Symbols

Abbreviations, upper case

3D	three-dimensional
AC	alternate current
AlN	aluminium nitride
AM	amplitude modulation
ASE	amplified spontaneous emission
AWGN	additive white Gaussian noise
BB	baseband
BER	bit error rate
BPSK	binary phase shift keying
BS	base station
CATV	community access television
CEPT	European Conference of Postal and Telecommunications Administrations
CNR	carrier-to-noise ratio
CPE	common phase error
CPW	coplanar waveguide
CS	central station
CSO	composite second order
CTB	composite triple beat
Cu	copper
CW	continuous wave
DBR	distributed Bragg reflector
DC	direct current
DD	direct detection

DFB	distributed feedback
DR	dynamic range
DSB	double sideband
E/O	electro-optical
EAM	electro-absorption modulator
ECMA	European Computer Manufacturers Association
EDFA	erbium-doped fiber amplifier
EIRP	equivalent isotropic radiated power
EOM	electro-optical modulator
ESA	electrical spectrum analzyer
ETSI	European Telecommunications Standards Institute
EVM	error vector magnitude
FBG	fiber Bragg grating
FFT	fast Fourier transformation
FM	frequency modulation
FOM	figure of merit
FP	Fabry-Pérot
FSR	free spectral range
FWHM	full width at half maximum
Gbps	gigabit per second
Ge	germanium
GHz	gigahertz
GN	Gaussian noise
GSM	Global System for Mobile Communications, originally Groupe Spécial Mobile
GVD	group velocity dispersion
HD	high definition
HDMI	high definition multimedia interface
HPA	high power amplifier
Hz	Hertz
I/Q	inphase-quadrature
ICI	inter-carrier interference
IEEE	Institute of Electrical and Electronics Engineers
IF	intermediate frequency

IFoF	intermediate-frequency-over-fiber, IF-over-fiber
IIP2	second-order intercept point referred to input power
IIP3	third-order intercept point referred to input power
IM	intensity modulation
InGaAs	indium gallium arsenide
InGaAsP	indium gallium arsenide phosphide
InP	indium phosphide
IP1dB	1 dB compression point referred to input power
ITU	International Telecommunications Union
kHz	kilohertz
$LiNbO_3$	lithium niobate
LO	local oscillator
LTI	linear time-invariant
TL	translation loop
MAC	media access control
MB-OFDM	multiband orthogonal frequency division multiplexing
Mbps	megabit per second
MHz	megahertz
ML	mode-locking, mode-locked
MLLD	mode-locked laser diode
MMF	multi mode fiber
MT	mobile terminal
MZM	Mach-Zehnder modulator
NTC	negative temperature coefficient
O/E	opto-electrical
ODSB	optical double-sideband
ODSB-SC	optical double-sideband with suppressed carrier
OFDM	orthogonal frequency division multiplex
OIP2	second-order intercept point referred to output power
OIP3	third-order intercept point referred to output power
OP1dB	1 dB compression point referred to output power
OSA	optical spectrum analzyer
OSSB	optical single-sideband
OTDM	optical time division multiplex

PD	photo-detector
PFD	phase frequency detector
PHY	physical layer
PLL	phase-locked loop
PM	phase modulation
QAM	quadrature amplitude modulation
QPSK	quadrature phase shift keying
RAP	remote access point
RBW	resolution bandwidth
RF	radio frequency
RHD	remote heterodyning
RIN	relative intensity noise
RoF	radio-over-fiber, RF-over-fiber
Rx	receive
S	sulfur
SA	saturable absorber
SC	single-carrier
SFDR	spurious-free dynamic range
SMF	single mode fiber
SNR	signal-to-noise ratio
SOI	second order intermodulation
SSB	single sideband
TEM	transmission electron microscopy
THz	terahertz
TL	translation loop
TOI	third order intermodulation
Tx	transmit
UTC	uni-travelling carrier
UWB	ultra wideband
VCO	voltage controlled oscillator
VCSEL	vertical-cavity surface-emitting laser
VOA	variable optical attenuator
VSA	vector signal analyzer
WDM	wavelength division multiplex

WLAN	wireless local area network
WPAN	wireless personal area network

Abbreviations, lower case

μ-wave	microwave wave
laser	light amplication by stimulated emission of radiation
mm-wave	millimeter wave
p-i-n	p-doped, intrinsic, n-doped
pdf	probability density function
psd	power spectral density
rms	root-mean-square

Constants

ε_0	electric constant, vacuum permittivity: $8.854187817620 \cdot 10^{-12}$ F/m
c_0	speed of light; in vacuum: 299792458 m/s
e	Euler number: 2.7182818284
h	Planck constant: $6.6260695729 \cdot 10^{-34}$ Js
k_B	Boltzmann constant: $1.380648813 \cdot 10^{-23}$ J/K
q	charge of an electron: $1.60217646 \cdot 10^{-19}$ C

Greek symbols

α	attenuation constant of the fiber
α_H	linewidth enhancement factor
α_m	modulator chirp value
β	FM index
β_L	linear part of the complex propagation constant
β_NL	non-linear part of the complex propagation constant
β_p	complex propagation constant
$\Delta v_{\mathrm{F},m}$	linewidth of an electrical beatnote
Δv_m	linewidth of an optical mode with mode number m

Δf_r	resonator bandwidth
η_L	laser quantum efficiency
Γ_L	reflection coefficient of the laser chip
κ_0	ratio of fiber and laser mode field radii
λ	optical wavelength
ν	optical frequency
ν_F	pulse repition frequency, MLLD self-pulsation frequency, free spectral range
ν_m	optical mode frequencies with mode number m
ϕ_k	phase of photocurrent of the kth harmonic of the fundamental beat signal
ρ	responsivity of the photo-detector
$\sigma^2_\mathrm{ASE-ASE}$	ASE to ASE beat noise power
σ^2_d	detected noise power
σ^2_RIN	RIN power
$\sigma^2_\mathrm{sig-ASE}$	signal to ASE beat noise power
σ^2_sn	shot noise power
σ^2_th	thermal noise power
σ_rms	integrated rms phase error
φ_IF	phase of IF signal
φ_LO	phase of LO signal
φ_m	phase of optical mode with mode number m
φ_ref	reference phase
φ_RF	phase of RF signal
ζ	noise process parameter
f	electrical frequency, Fourier variable

Latin Symbols

A	effective detection area of a photo-detector
B_o	optical bandwidth
c	speed of light
D	dispersion parameter

e	amplitude error
$e_{\mathrm{opt}}(z,t)$	electrical field component of an optical wave
$e_{\mathrm{RF}}(z,t)$	RF envelope of an optical wave
$E_{\mathrm{TxRx}} = 0$	deterministic error introduced by receiver or transmitter imperfections
$EVM_{\mathrm{rms,ap}}$	rms EVM normalized to average symbol power
$EVM_{\mathrm{rms,pp}}$	rms EVM normalized to peak symbol power
F	linear noise figure
f_{fc}	flicker corner frequency
f_{IF}	intermediate frequency
f_{LO}	local oscillator frequency
f_{m}	modulation frequency
f_{RF}	radio frequency
f_{r}	resonator frequency, relaxation frequency
g	link gain
$G_{\mathrm{e}}(v)$	EDFA gain spectrum
G_{F}	forward transfer function
g_{i}	intrinsic link gain
G_{R}	reverse transfer function
H	closed-loop transfer function
H_{m}	modulator transfer function
H_{SMF}	transfer function of SMF
I	optical intensity
I_{B}	laser current bias point
I_{k}	amplitude of photocurrent of the kth harmonic of the fundamental beat signal
i_{ph}	photo current
I_{P}	charge pump current
k	wave vector
K_{a}	amplifier gain
K_{IQ}	insertion loss of an I/Q modulator
K_{m}	mixer conversion loss
K_{P}	charge pump gain
K_{V}	VCO tuning sensitivity

l	fiber length
$L(\kappa_0)$	mode field mismatch
$L_\varphi(f)$	single-sideband power spectral density of phase φ
L_{ar}	passband attenuation of a band-pass filter
L_A	stopband attenuation of a band-pass filter
l_{cc}	chirp-induced offset of the first power maximum in a dispersive environment
L_{CRITICAL}	usable fiber length around a power maximum
l_p	periodicity of power maxima in a dispersive environment
L_{WD}	split ratio of a Wilkinson divider
M	number of modes in the MLLD spectrum
$m(t)$	field modulation function
m_i	IM index
M_p	number of polarization modes
N	divider ratio of a frequency divider
n	refractive index
N_{dsc}	number of data sub-carriers in an OFDM transmission
N_p	number of frames in an OFDM transmission
N_s	number of symbols in an OFDM transmission
NF	logarithmic noise figure
P_{avg}	average signal power
$p_{\mathrm{d,opt}}$	detected optical power
$p_{\mathrm{g,a}}$	available power from the generator
p_{in}	input power
p_{load}	power delivered to the load
$p_{\mathrm{m,opt}}$	optical modulation power
p_{opt}	optical power
p_{out}	output power
$p_{\mathrm{RF,c}}$	electrical power in the mm-wave carrier
p_{RF}	electrical power
PAV	peak-to-average power ratio
Q	oscillator quality factor
Q	quality factor
R	divider ratio of a frequency divider

R_0	standard load impedance
R_i	ith normalized received symbol
s	Laplace variable
$S_\varphi(f)$	power spectral density of phase φ
S_i	ith normalized ideal symbol
s_l	slope efficiency
T_0	reference temperature: 290 K
T_b	fraction of recovered RF power in one of the fundamental sidebands
T_{cap}	capture time of the PLL
T_i	MZM fiber-to-fiber transmission
T_L	optical transmission loss
V_π	MZM halfwave voltage
v_φ	phase velocity
V_B	MZM voltage bias point
v_g	group velocity
w	mode field radius
X	input variable of an LTI system
Y	output variable of an LTI system
Z_{eqLPF}	equivalent low-pass filter transfer function
Z_{LPF}	loop filter transfer function

List of Tables

1 Millimeter-Wave Radio Communications across the Analog Optical Link

In a communication system, information is transmitted from one place to another by means of an electromagnetic carrier wave whose frequency might lie in the range of a few megahertz up to several hundreds of terahertz, depending on the properties of the channel used for transmission. The rate at which information can be sent across a transmission channel is theoretically limited by the channel capacity which, in turn, scales with the bandwidth available to the communication system[1] [Sha49].

Future networks will be required to provide data rates up to several tens of gigabit per second in order to cope with the exigencies of such diverse markets as consumer electronics and healthcare. The key applications driving much of today's demand are the following ([YXVG11], [Wir10]): uncompressed high definition (HD) video streaming with a performance at least equivalent to wired displays, support of 3D video formats, wireless gaming featuring very short periods of latency, high-speed file transfer between wireless docking stations, and, in the realm of telemedecine, health monitoring and real-time active services. Thus, the demand for data rate directly translates into a demand for bandwidth. Other than speed, another principal feature of modern communication networks is maximum possible user comfort, ensured by the wireless access to any local or global subscriber service. The link must then be terminated with an antenna by means of which the signal is transmitted across the air interface to one of many wireless terminals in the form of electromagnetic radio waves. In a home or office environment where such networks (referred to as *wireless personal area networks*,

---

[1]In a channel limited by additive white Gaussian noise, $C = B \log_2 SNR$, where C is the channel capacity, B is the bandwidth, and SNR is the signal-to-noise ratio.

WPAN) will primarily be employed, the distances to be bridged by the wireless link are short, i.e. in the range of a couple of meters.

Wireless communications, however, bring about their very own set of constraints on bandwidth, most notably, the strongly regulated use of frequency bands. Conventional wireless systems (e.g. IEEE 801.11x) operate in the microwave region, where bandwidth is extremely limited[2]. There are however regions in the electromagnetic spectrum where this contraint is relieved: the wordwide availability of at least 5 GHz unlicensed bandwidth in the millimeter wave region around 60 GHz has drawn a lot of attention to this particular band. Gigabits per second data rates are feasible when multi-carrier techniques are used [Smu02]. The high free-space loss[3] these frequencies imply is *de facto* uncritical in a WPAN, but for area-wide signal distribution, a large number of access points are required. If baseband-to-millimeter wave conversion was to be performed directly at the antenna, overall system costs would be driven up by the sheer multitude of access points. A considerable cost benefit could be gained by centralizing some of the signal processing and by distributing a "ready-to-radiate" analog signal to the access point. Due to the cable losses in the millimeter wave range[4], electrical feeder networks are not power-efficient.

In an *optical* communication system, lightwaves in the near-infrared region (approx. 200 THz) serve as carrier signals. Due to the phenomenon of total internal reflection, they can conveniently be guided by doped silica fibers. Information can be modulated onto the lightwave using either a digital (discrete with time) or an analog (continuous with time) scheme. While the progress and enormous success of optical telecommunications over the last decades have largely relied on digital links, analog links have been under intense study for a growing number of applications ([CI04], [Koo06]). Both analog and digital optical links profit from the key features of the fiber as a highly flexible transmission medium characterized by low loss and large bandwidth. At wavelengths in the so-called "telecom window" around 1.55 μm, a signal can travel as far as 40 km across an optical link

---

[2]IEEE 801.11x uses the 2.4 GHz or 5 GHz bands, respectively, with a channel bandwidth of 20 to max. 40 MHz.

[3]The free-space loss is proportional to the square of the frequency f^2, i.e., from 6 GHz to 60 GHz it rises by 20 dB.

[4]An attenuation of about 5 to 10 dB per meter is not unusual for coaxial cable at 60 GHz.

before it needs amplification, whereas an amplifier is usually needed each 3 to 4 km on a copper line. The virtually unlimited bandwidth of the optical fiber has enabled a large number of applications relying on high-speed data transmission through the networks which span the globe.

If the optical link terminates in an electrical wireless link, as in the case of the WPAN, it can advantageously be implemented as an analog link interfacing directly with the antenna unit, resulting in a simpler and potentially more cost-effective architecture.

Thanks to its characteristic features, the fiber can accommodate low-loss transmission of millimeter wave signals in a feeder network where the signals are optically distributed. The challenge then lies with finding the appropriate means for electro-optical and opto-electrical conversion of those signals at the transmitting and receiving ends of the link. Typically, a converter pair consisting of a laser and a photo-detector is employed to modulate and demodulate the electrical signal on and off the lightwave carrier. Naturally, this double conversion is only justified when the conversion process does not itself introduce a drawback in terms of the two key features rendering the optical fiber attractive in the first place: bandwidth and loss. Furthermore, it is desirable to keep the overall implementation expenditure at an absolute minimum for broad market acceptance.

While it must be recognized that the concept of analog fiber links has been under study for a considerable amount of time - notably for community access television (CATV) systems [DB90]-, and although several commercial products like Litenna^{TM} or BriteCell^{TM} operating in the microwave range have been brought to market[5], the field of microwave photonics has not yet delivered a very large scale application to the end consumer.

It is very probable that the potential of analog optical fiber links will now be fully realized in the millimeter wave range. At the same time, major involvement of the industrial global players and large scale commercialization will only set in when device production and the implementation of link architectures become low-cost enough to allow for the laws of economies of scale to apply - especially in the consumer electronics market.

---

[5]Tekmar's BriteCell^{TM} system permitted the flexible installation of more than 500 optically-fed remote antennas for the dynamic allocation of GSM network capacity for the visitors of the 2000 Olympic Games in Sydney, Australia.

In the following sections, the implications of millimeter wave wireless communications will be reviewed, as well as the ongoing regulation and standardization effort related to the 60 GHz range. The basic concepts currently employed for the transmission of radio frequency signals across the fiber optic link will be revisited. From this preliminary analysis, it shall become apparent that the current state of the art does not yet encompass a technique which solves the problem of the remote delivery of modulated millimeter wave signals in a satisfactory way. This is mainly due to the fact that little work has yet been undertaken aimed at the integration of components or whole systems.

At the heart of this work lies a component which allows the system designer to move one step closer to a fully integrated system: the mode-locked laser diode. This miniature device provides electro-optical conversion and up-conversion to the millimeter wave range in a simple configuration and thus has the potential to not only render analog fiber optic links more cost-effective, but also to replace electrical millimeter wave oscillators wherever needed when combined with an appropriate photodiode in the same package. The properties of millimeter wave generation based on this device, and its system-level implications, will be the focus of this dissertation.

1.1 Communications in the 60 GHz region

In the broader sense, millimeter wave (mm-wave) technology is concerned with the part of the electromagnetic spectrum between 30 and 300 GHz, corresponding to wavelengths from 10 mm to 1 mm [Poz05]. In the context of this work, the focus will be on the region around 60 GHz, as it is of particular interest for the type of application regarded in this work[6].

Though known for many decades [Olv89], it has only been through the progress in silicon process technologies and cost-efficient integration in the past couple of years that mm-wave technology has become relevant beyond military applications [YXVG11]. The increased path loss at mm-wave frequencies as well as the high atmospheric attenuation due to the oxygen resonance at 60 GHz

---

[6]In the following, the expression "60 GHz transmission" and "mm-wave transmission" will be used as synonyms for mm-wave transmission in the band from 57 - 66 GHz.

have hindered mid- to long-range radio communications in this band and limited free-space mm-wave transmission to a few meters even in line-of-sight condition. In an indoor environment, frequency-related path loss is considered the dominant loss contribution [Yon11]. Although usually considered a drawback in radio communications, there are, however, applications that can benefit from this limited coverage, such as high data rate indoor WPANs: The restricted coverage of a WPAN facilitates frequency reallocation, and thus allows for high user-densities in a picocell architecture. Another important aspect of WPANs operating in the 60 GHz range is that they provide inherent network security as electromagnetic waves will be considerably attenuated by walls which represent reliable cell boundaries [Smu02].

Another aspect of 60 GHz radio communications is to be evaluated in the context of the environmental consequences implied by the indoor communication system penetrating our anthroposphere. In a home environment, rooms not intended for network access remain entirely field-free. 60 GHz radiation can furthermore be considered a reduced risk, because the depth of human tissue penetrated by measurable quantities of electromagnetic energy amounts to only a couple of tenths of a millimeter, depending on the underlying cutaneous structure which so shields the inner organs [ARLZ08].

1.1.1 Regulation and standard development in the 60 GHz region

Large bandwidth being the most promising feature of mm-wave radio communications, it is worthwhile to have a look at how regulatory bodies worldwide have allocated the unlicensed band around 60 GHz. At least 5 GHz of continuous bandwidth is available in many countries [YXVG11].

In particular, the European Telecommunications Standards Institute (ETSI) and the European Conference of Postal and Telecommunications Administrations (CEPT) are working towards a legal framework for the deployment of unlicensed 60 GHz devices in Europe [Ins]. Under the current proposal of the ETSI, 9 GHz are allocated between 57 GHz and 66 GHz. This band regroups the bands currently approved among major European countries. The reader is referred to [Off] for up-to-date information on European regulation. For other regions, the following sources might be of interest: [Com04] and [MT05] for North America, [oPMHA00] for Japan, or [CA05] for Australia.

Compared to the similarly large unlicensed bandwidth allocated for over-lay ultra-wideband (UWB) technology (480 Mbps at an output equivalent isotropic radiated power (EIRP) of -10 dBm, and a power spectral density mask limited to -41.3 dBm/MHz), the power limits are much less restricted (40 dBm output EIRP[7]), and target bit rates about ten times higher (4 - 5 Gbps). The abundant bandwidth available for mm-wave radio communications also simplifies the design of such systems, as only low spectral efficiency is required, implying low system and implementation cost. In Europe, the maximum allowed transmission power at the antenna is 13 dBm [YXVG11]. Under current recommendations, a maximum EIRP of 40 dBm with a maximum spectral power density of 13 dBm/MHz for indoor applications is allowed [ERC11]. Earlier recommendations have limited the maximum antenna gain to 30 dBi [oPA09].

The strong commercial interest in 60 GHz systems has triggered considerable standardization work in parallel with the regulation effort. The European Computer Manufacturers Association (ECMA) published the first ratified version of the ECMA 387 specification for high-rate 60 GHz PHY, MAC and HDMI PAL in December 2008 [Int08]. In September 2009, the IEEE ratified the IEEE 802.15.3c standard as an alternative WPAN physical layer for the existing WPAN standard [IEE09] supporting data rates of at least 2 Gbps for short-range wireless transmission [IEE09]. Furthermore, the IEEE is to publish an amendment to the IEEE 802.11-2007 standard under the suffix -11ad, enabling very high throughput at 60 GHz. The ratification was expected for late 2012 [IEE12]. This standard is meant to coexist with other 60 GHz standards and aims at 1 Gbps transmission while maintaining the network architecture of a conventional IEEE 802.11 system as well as full backward compatibility. Virtually all of the standards employ digital modulation and both single carrier (SC) and multi-carrier formats, notably orthogonal frequency division multiplexing (OFDM) techniques.

---

[7]Indoor, in Europe. Maximum output EIRP in the USA is 43 dBm.

The Wireless HD consortium[8] and the Wireless Gigabit Alliance[9] are equally working towards a quick commercialization of 60 GHz systems under unified specifications which can serve a large number of applications ([Wir10], [All10]).

Conclusion: The standardization effort reflects the immense interest in 60 GHz systems as effort and resources of various global industrial players are pooled in order to come up with a powerful and, above all, sustainable wireless ecosystem.

1.1.2 Convergence of fiber-wireless systems

The free-space loss at 60 GHz is, in general, beneficial to small cell sizes. Within the cell, it can be overcome by making proper link budget considerations and using highly directive antennas. However, mm-wave physics imply yet another difficulty for the link designer: Due to the skin effect, a 60 GHz wave will experience significant loss when propagating through a coaxial waveguide. A common value is an attenuation of 5 to 10 dB per meter [TEC12]. Thus, it seems impractical to feed a potentially very large number of remote antenna points (RAPs), e.g. in a home or office environment, through a coaxial network. Yet full processing of baseband data at each base station (BS), as in the case of a type of WLAN router with extended mm-wave capabilities, results in complex BS and cannot be favored for large networks either.

It is thus intuitive to combine the high bandwidth mm-wave radio systems with the bandwidth-generous optical fibers[10], ideally achieving perfect coexistence of optical baseband (digital) transmission and optical mm-wave transmission at reasonable loss. To that end, system designers have gained valuable experience in the past, most notably with CATV technologies [DB90]. A general tendency of convergence between fiber links and wireless links has been acknowledged [KYJ+08]. The fiber-wireless link is sometimes referred to as *hybrid* link

---

[8]Including Broadcom Corporation, Intel Corporation, LG Electronics, Panasonic Corporation, Royal Philips Electronics, Samsung Electronics, Sony Corporation, Toshiba Corporation, and others. First devices have been brought to market.

[9]Including AMD Corporation, Microsoft Corporation, Dell Corporation, Intel Corporation, Panasonic Corporation, Samsung Electronics, Nokia Corporation, Toshiba Corporation, and others.

[10]Modern fiber equipment is capable of speeds in the order of tens of terabits per second baseband transmission.

[GSLO99], and is such in at least two respects: It is hybrid as it allows for wave-guide and wireless transmission, and it is hybrid in the sense that both electrical and optical signals are processed at distinct points of the link.

1.1.2.1 Analog schemes for the optical transmission of data signals

The conventional approach to the distribution of data across the fiber link is the baseband modulation of a light source in order to transmit digital data. This technique is known as *baseband-over-fiber* scenario and represents the most common type of digital optical communication both downlink (i.e. traffic from the network to the user) and uplink (i.e. traffic from the user to the network). Where a hybrid link including wireless transmission is concerned, this approach requires the deployment of signal processing units to the RAP or BS, such as digital-to-analog-converters, I/Q-modulators and the like. In a scenario where a large number of RAPs need to be fed, this configuration quickly drives up overall costs, which is why a less complex setup is preferred. Using an analog transmission scheme, as shown in figure 1.1 (downlink), the digital baseband (BB) data stream is processed in the central station (CS). It is modulated onto an analog waveform, typically by a vector modulator, and, in turn, this analog waveform of continuous amplitudes at radio frequency (RF) or at an intermediate frequency (IF) is then used to modulate the light source. Thus, the expression "analog modulation" is justified, even if the BB modulation format into the link is a digital format (such as BPSK, QAM), suggesting a set of discrete available amplitudes[11].

Electro-optical conversion of the analog signal is typically achieved by modulating a laser diode. The optical signal is then sent across the fiber and back-converted to the electrical domain by a photo-detector, typically a photodiode, from which it is routed directly to the antenna and sent across the wireless interface to the mobile terminal (MT). All system intelligence and complexity is thus located at the CS, rendering the BS, at least theoretically, passive. In practice, amplifiers and filter devices might have to be included in the BS.

This approach encompasses both the *IF-over-fiber* (IFoF) and *RF-over-fiber* (RoF)[12]

---

[11]Compared to digital modulation, where the modulation index is ideally equal to one (the laser is ideally turned off and on completely), the modulation index for analog optical modulation is low.

[12]In the context of this work, "RoF" refers to the transmission of mm-wave signals across the fiber, as

techniques, the difference between the two being an additional mixing unit at the BS for IFoF [MIMH08]. While the advantages of the analog techniques seem obvious, analog modulation usually imposes more restrictive requirements on the light source, such as linearity to prevent distortion, high conversion efficiency, and low noise performance in order to limit the link noise figure [ACIR98].

1.1.2.2 Architecture and requirements definition

The decision for an IFoF or a RoF transmission scheme must be made with regard to the requirements on cost and complexity. The IFoF/RoF link is typically a short-haul to medium-haul point-to-point or point-to-multipoint link required for full duplex operation [GMA09]. In-house transmission distances are in the order of a couple of hundred meters. A point-to-point architecture between a central station and a mobile terminal across a base node is assumed in the following. It is important to note that the downlink (CS to BS to MT) most probably accounts for the better part of the traffic: a user is more likely to download large amounts of data (e.g. in streaming applications) than to upload them. The primary concern is therefore to provide a mm-wave downlink:

The system will then profit from a full-fledged RoF transmission (see figure 1.1, downlink): the analog mm-wave is sent out from the CS and transported all the way to the MT. The base station comprises two main components: an opto-electrical converter and the antenna module which transmits the wireless signal. In all cases, the electro-optical converter must operate in the mm-wave range. Due to the scarcity of such components, this very fact represents the key challenge in RoF system design.

For the uplink (MT to BS to CS), signal transmission in the inverse direction must be enabled. Here, an IFoF scheme can be chosen. This will result in a more cost-effective design, given the abundance of low-cost laser devices with modulation bandwidths in the IF range. If we assume that the uplink traffic does not actually require very high data rates, the wireless uplink can be realized using the conventional wireless LAN standard (IEEE 802.11x [IEE99]) operating at microwave (μ-wave) frequencies (see figure 1.1, uplink). Electro-optical conversion in the μ-wave range is possible without difficulty. We can even imagine

---

compared to IF transmission at, e.g., microwave frequencies (hundreds of MHz to several GHz).

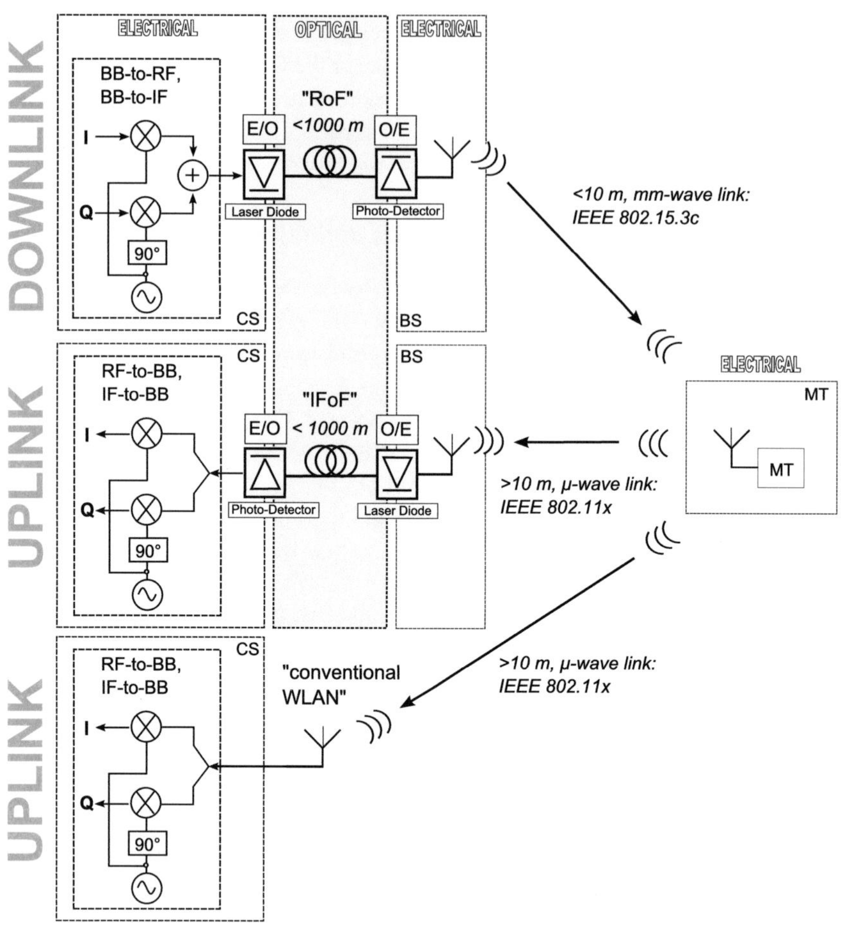

Figure 1.1: Possible downlink and uplink configurations

a scenario where the uplink is implemented as a conventional wireless LAN up-link from the MT to a central router. Here, the mm-wave RoF downlink basically represents a high-speed extension to an existing system in the case of highly asymmetrical traffic.

Alternatively, the uplink could also operate at mm-wave frequencies. The question then remains as to how the received mm-wave signal can be translated to the IF range without including another mm-wave local oscillator at the BS. This problem has been solved by bandpass-filtering the mm-wave carrier in the downlink path and routing it to the mixer (down-converter) on the uplink path [LKON06]. It is even imaginable to make use of the non-linear characteristics of the opto-electrical converter, i.e. of the photodiode, to perform the mixing operation [PGMA98].

We maintain that in the duplex system, the mm-wave downlink is the more critical of the two directions. It shall therefore be the focus of this work. For the downlink architecture, certain specifications are required regarding the performance of the analog optical feeder link in terms of power and dynamic range. In order to exploit the maximum allowed EIRP, it is preferable that the feeder network incorporating the analog optical link be capable of delivering an amount of power approaching the regulated maximum. Yet the electrical output power of a RoF link is always limited by the inherently inefficient double-conversion between the electrical and the optical domain. For this reason, electrical high power amplifiers (HPA) have been a common part of the link. Finally, in terms of dynamic range, target figures in the range of 77 - 80 dB· $\text{Hz}^{2/3}$ have been stated for various scenarios, where it was assumed that third order distortion was the limiting factor [FLK97].

Conclusion: For all the above-mentioned reasons, we establish the analog fiber-wireless link as the preferred solution for mm-wave feeder networks for the scope of this work. It is acknowledged that, in general, any infrastructure that will ultimately terminate in a wireless transmission link can benefit from an analog technique. The downlink requires mm-wave generation and modulation capabilities and thus represents a bigger challenge compared to the uplink which can be realized at μ-wave frequencies. A RoF technique aimed at maximum transmission power at optimum dynamic range shall be used, and the link will need to cover distances of several hundred meters.

1.2 The RF-over-fiber link: State of the art

In the following, basic concepts of the design of radio-over-fiber links are reviewed, and state-of-the-art techniques for optical mm-wave generation and modulation, as well as their detection, are revisited.

1.2.1 Components of a radio-over-fiber link

The RoF link comprises three basic functions (figure 1.1), namely electro-optical (E/O) conversion at the transmitter, signal transmission across the fiber, and opto-electrical (O/E) conversion at the receiver. Whether or not a pair of E/O-, O/E-converters is suitable for mm-wave transmission strongly depends on their respective performance in terms of modulation and detection bandwidths. While the transmission loss across the fiber is inherently small even for larger distances, the *zero link* loss induced by the converters (E/O, O/E) is one of the crucial elements of RoF link design. The figures of merit used in this context are the modulation and detection efficiencies at the interfaces of the electrical and the optical domain. Typical zero link losses range from 30 to 40 dB [CI04]. While these values appear relatively large and need to be accounted for in the link budget, this circumstance should not obscure the fact that the total resulting loss is almost perfectly independent from link length which clearly is a conceptual advantage over electrical mm-wave links.

1.2.1.1 Fiber transmission

The optical fiber consists of a core of silica glass surrounded by a cladding whose refractive index is lower than that of the core. Thus, light can be guided by total internal reflection. The propagation of optical fields in fibers is governed by Maxwell's equations which result in the fiber's wave equation. The notion of a *mode* then relates to a particular solution of the wave equation which satisfies the appropriate boundary conditions and which has the property that its spatial distribution does not change with propagation. Like in electrical waveguides, the boundary conditions are given by the waveguide material and the geometry of the fiber. *Single-mode fiber* (SMF) has a core radius comparable to the light wavelength (5-8 µm). It guides only the fundamental mode HE_{11}. *Multi-mode fiber* (MMF) has a core radius of 50-62 µm and guides multiple modes.

The fiber is characterized by two principal effects: attenuation due to absorption in the material and dispersion. While modal dispersion can be avoided using SMF, other dispersion effects are not so easily steered clear of. Even in SMF, waveguide dispersion due to imperfections of the refractive index profile can play a role. Chromatic dispersion refers to the different speeds at which signals at different optical frequencies v propagate due to the change in refractive index of silica with v. For silica glass, the dispersion minimum is found at about 1.31 µm, but dispersion-shifted fibers can be fabricated for which attenuation and dispersion are minimized at 1.55 µm [Agr97]. While polarization-mode dispersion is present in SMF, its effects are negligible compared to those of chromatic dispersion for relatively short transmission distances < 1 km. Likewise, non-linear effects, like four-wave mixing or inelastic (Raman or Brillouin) scattering, are negligible for low power levels (< 10 mW) in the fiber, and such is the case for the analog fiber optic links presented herein.

The loss caused by the fiber's attenuation is wavelength-dependent; its practical limit was found at 0.2 dB/km at a wavelength of 1.55 µm [MTHM79], which corresponds to the optical C-band. In the C-band, optical amplifiers based on rare earth material are available, which is why this wavelength range has been favored for telecommunications since the 1980s (*telecom window*). Consequently, laser sources and photo-detectors have also been developed for this wavelength range.

1.2.1.2 Electro-optical conversion at the transmitter

The only light source to date which is of practical interest for the target application is the laser[13] diode for its increased gain and modulation bandwidth, beam coherency and increased fiber coupling efficiency compared to e.g. the light emitting diode. Its operation is based on the phenomena of stimulated emission and radiative recombination of charge carriers at an intraband transition in the excited laser medium, and on a coherent feedback system forming a resonating structure. The choice of semiconductor material determines the emitted wavelength. For an emission around 1.55 µm, the compound semiconductor indium-

---

[13] light amplification by stimulated emission of radiation

gallium-arsenide-phosphite (InGaAsP)-indium-phosphite (InP) material system can be chosen.

The charge carrier dynamics of the laser material as well as the laser architecture determine the achievable modulation bandwidth. The intrinsic modulation bandwidth of current-pumped *single-chip* laser diodes does not yet extend to the full mm-wave domain. In distributed feedback lasers (DFB), modulation bandwidths up to 35 GHz have been shown using integrated feedback techniques [KVT$^+$11]. Vertical-cavity surface-emitting lasers (VCSELs) also exhibit inherently large bandwidths up to several tens of GHz [HMW$^+$11]. Optically injection-locked VCSELs have been reported to exhibit bandwidths up to 80 GHz [LZS$^+$08], which come however at the expense of complex master-slave laser architectures.

1.2.1.3 Opto-electrical conversion at the receiver

From an electrical point of view, the photo-detector performing opto-electrical conversion is a diode in reverse bias[14]. When illuminated, the diode conveys electrical carriers generated through an absorption process to an external circuit. The photocurrent can usually be considered proportional to the optical power incident on the diode. As the intensity of an optical wave is proportional to the squared magnitude of its electrical field $\left|e_{\mathrm{opt}}\right|^2$, the diode represents a square law detector. The diode material must be chosen such that it can absorb light at the wavelengths of interest. Germanium (Ge) and the compound semiconductor indium-gallium-arsenide (InGaAs) have bandgap energies that correspond to wavelengths around 1.6 μm, and both materials can be used for the bands at 1.3 μm and 1.55 μm. InGaAs diodes are more frequently used, as they exhibit lower dark currents than Ge diodes [CI04]. A favored photo-diode structure has been the p-doped - intrinsic - n-doped (p-i-n) diode [AM97] for its low noise performance at 1.55 μm compared to e.g. avalanche photodiodes [HKM77], although uni-traveling-carrier (UTC) diodes have recently gained much attention for very high-speed applications [SAH$^+$09].

Present-day commercial p-i-n photodiodes reach bandwidths of up to 110 GHz. In general, UTC photodiodes provide higher operating frequencies than conven-

---

[14]Forward biasing the structure would cause an electrical current to flow that would largely exceed the photo-generated current.

tional p-i-n photodiodes; they have been shown to operate up to 300 GHz [NII09]. Details on photo-detection can also be found in appendix A.

Conclusion: In all RoF systems presented herein, single-mode fiber is used to avoid effects of modal dispersion. The operating wavelengths center around 1.55 µm, as numerous components, such as optical amplifiers or photodiodes, are optimized for operation in this wavelength range due the low attenuation characteristics of silica fiber. The injection current into a laser diode can be modulated in order to achieve RF modulation of light. To date, single-chip diode lasers do not reach modulation bandwidths that extend to the 60 GHz range. Photodetectors are readily available in the 60 GHz range.

1.2.2 Modulation of light

Any type of communication system can be classified according to the type of modulation used, i.e., the electrical or optical variable varied according to the information signal to be transmitted. The choice of modulation format at the transmitter obviously requires an appropriate receiver architecture. For lightwave systems, two principal types of modulation are used: Field modulation and intensity modulation.

1.2.2.1 Field modulation and coherent detection

Whenever one of the characterizing parameters of an optical wave - amplitude, frequency or phase - is varied proportionally to a modulation signal, we speak of field modulation. Optical field modulation techniques can be regarded as direct extensions of the techniques that have been known for a long time in the electrical microwave domain. Unfortunately, field modulation techniques are rather difficult to implement in the context of optical systems. In a minimum configuration, at least three components are necessary: an optical local oscillator, an optical mixer (e.g., the photodiode [MC07]), and an appropriate filter device. One is deprived of the possibility of direct modulation, see section 1.2.2.2, as the direct variation of the frequency or the phase of the optical signal is difficult. On the transmitter side, highly coherent lasers are necessary, and on the receiver side, coherent detectors have to be used which tend to be complex. The hardware effort might well be worthwhile as the improvement of receiver sensitivity might yield up to 20 dB, and, as a result, transmission distances can be increased [Agr97]. As in electrical microwave systems, a more efficient use of bandwidth is possible. On the other hand, the often unknown state of polarization due to fiber birefringence[15] is usually a problem [Agr97], and can only be mitigated by the use of dedicated fiber material, polarization monitoring, or the employment of a polarization-insensitive coherent receiver.

1.2.2.2 Intensity modulation and direct detection

In intensity modulation (IM) systems, the intensity of the transmitted optical signal is proportional to the modulation signal, and the modulation can be recovered in a simple way by the use of a direct-detection (DD) receiver[16]. A photodiode as described in appendix A can be employed.

Compared to coherent systems, IM systems employing direct detectors achieve limited sensitivity, but the availability of optical pre-amplifiers has made it possible to approach the performance of systems employing coherent receivers [Agr97]. On the other hand, IM systems present the designer with the advantage of direct laser modulation, where the laser light is modulated through a variation of the injection current into the laser in a simple setup. Furthermore, IM systems are not sensitive to different polarization states. Indeed, the universal choice for RoF applications today is a technique combining intensity modulation of the optical carrier followed by a direct detection receiver [CI04]. We maintain that there is no urgent bandwidth constraint for IM, and the availability of an optical C-band amplifier permits the use of a simple direct detector.

Figure 1.2 shows the two basic schemes for E/O-conversion by intensity modulation. Using *direct modulation*, the electrical signal is directly applied to the

---

[15] If the fiber dimension deviates from the perfectly cylindrical core of uniform diameter, or if the fiber experiences non-uniform mechanical stress, the cylindrical symmetry of the fiber is broken. The fiber is said to have acquired birefringence, where the effective mode indices for the orthogonally polarized fiber modes differ from each other.

[16] It is thus important to note that amplitude modulation and intensity modulation are not identical, although they share some basic properties: For low modulation indices, their spectra have approximately the same bandwidth, otherwise the optical spectrum contains harmonics of the modulation waveform [CI04].

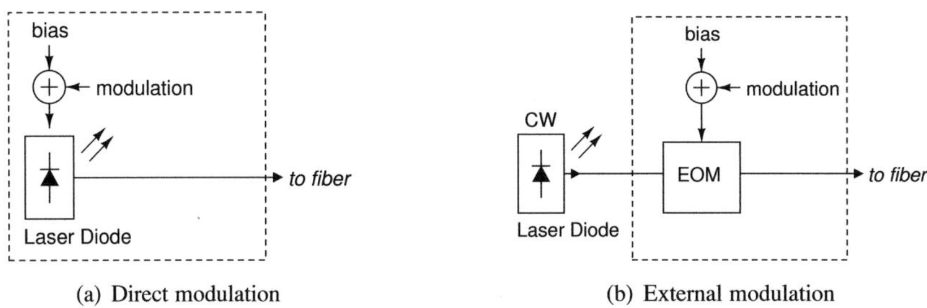

(a) Direct modulation   (b) External modulation

Figure 1.2: E/O conversion (dashed lines) using basic intensity modulation techniques

laser, and the output optical intensity changes with the current variation as long as the modulating signal is within the laser's bandwidth. The bandwidth constraint can be circumvented using *external modulation*, where the E/O converter comprises an electro-optical modulator (EOM), typically exhibiting a large bandwidth compared to the modulation bandwidth of the laser source providing the continuous-wave (CW) light signal. The laser then operates at constant optical power, and intensity modulation is achieved at the external modulator by applying the modulation signal to its electrodes. The most common type of external modulator is based on a Mach-Zehnder interferometer structure, making use of the so-called *Pockels effect*([CI04], [NCLG09], and appendix B). Modulators based on the electro-absorption effect in semiconductors have also been gaining increasing attention in recent years [PKS11].

Conclusion: Intensity modulation is a straightforward technique to impress an information signal onto a lightwave. Sensitivity issues can be avoided using amplifiers, so the advantages of coherent detection do not outweigh the relatively high complexity of such receivers. Consequently, this work will follow an IM-DD approach.

1.2.3 Optical signal generation

Among the various techniques that have been proposed for optical signal generation, any categorization one chooses to make is not obvious. We choose to classify techniques which rely either on a *single* optical carrier wave at one given wavelength modulated by the electrical signal, or techniques that rely on an optical transmitter whose output signal exhibits spectral components at *two or more* optical wavelengths. Using the latter, techniques of heterodyning can be employed, enabling signal generation at custom mm-wave frequencies.

1.2.3.1 Single wavelength approaches: Optical double sideband and optical single sideband

Optical double sideband (ODSB) transmission can be achieved either by directly modulating a laser's injection current around a bias point in the linear region of the laser's power-over-current transfer curve, or by using a linearly biased external modulator, e.g. a Mach-Zehnder modulator (MZM). The transfer function of the MZM is inherently non-linear, but it can be linearized around a bias voltage point where the slope of the curve is maximized. The reader is referred to sections 3.2.1.1 and to 3.2.1.2 for details. Both implementations result in a double sideband (DSB) spectrum centered around the optical carrier. For low modulation indices, this spectrum resembles an amplitude modulation (AM) spectrum in conventional microwave communications. The biggest drawback of this simple technique is the bandwidth limitation of the respective device. Laser diodes have large bandwidths compared to other laser architectures, but do not yet allow for direct modulation in the 60 GHz range, currently being limited to about 35 GHz [KVT+11]. External modulators achieve considerably higher modulation bandwidths. At this time, commercial modulators commonly reach up to 40 GHz[17]. ODSB transmission encounters the problem of electrical power drop at periodic link lengths due to chromatic dispersion. When the two detected sidebands combine at the photo-detector, they exhibit a phase difference which may result in destructive interference at certain link lengths. The problem can be overcome using techniques of *optical single sideband* (OSSB) transmission, where the mod-

---

[17]See for example, products offered by Gigoptix, JDSU, Photline, Thorlabs.

ulation is applied to the two electrodes of an MZM with a phaseshift of $\pi/2$ [SNA97]. It is, however, difficult to achieve a given phaseshift over a broad range of frequencies, which is why architectures employing fiber Bragg gratings have been employed to filter the undesired sideband [NDCL09]. In the 60 GHz range, an approach incorporating a directly modulated VCSEL has been proposed in [NFP$^+$10], where a VCSEL (slave) was injection-locked by means of a DFB (master) laser. An advantage of the injection-locked system is its robustness against dispersion effects thanks to single sideband (SSB) transmission; a clear disadvantage is, again, the relatively high complexity of the setup.

1.2.3.2 Multiple wavelength approaches: Remote heterodyning

In general, heterodyning refers to a technique where two signals of two distinct frequencies are mixed in order to give a signal exhibiting, typically among other mixing products, a third component at the sum or the difference frequency of the two original signals. In a receiver, the frequency v_2 of a local oscillator (LO) is often chosen to differ from the signal frequency v_1 such that a signal is formed at an intermediate frequency much lower than the signal carrier frequency.

In the context of a coherent analog fiber link, remote heterodyning (RHD) is employed in order to generate an electrical mm-wave signal by mixing the transmitted optical signal with an LO laser signal at the photo-detector. The principle of RHD can be illustrated by the interference at the detector of two optical wavelengths at the optical frequencies v_1 and v_2, represented by their respective electrical fields,

$$e_1(t) = A_1 \exp\left(2\pi v_1 t + \varphi_1\right) \tag{1.1}$$

and

$$e_2(t) = A_2 \exp\left(2\pi v_2 t + \varphi_2\right), \tag{1.2}$$

which results in an electrical photocurrent

$$i_{(t)\text{ph}} \propto (e_1(t) + e_2(t)) \cdot (e_1(t) + e_2(t))^*$$
$$= |e_1(t) + e_2(t)|^2 = |A_1|^2 + |A_2|^2 + 2A_1A_2 \cos\left(2\pi(v_2 - v_1)t + (\varphi_2 - \varphi_1)\right).$$
$$(1.3)$$

Electrical signals at arbitrary frequencies $v_2 - v_1$ in the mm-wave range can be created by choosing the lasers' wavelengths accordingly; for 60 GHz generation it is therefore sufficient to make sure that the two optical modes involved are separated by that frequency.

When the mm-wave signal is generated through a heterodyning technique, the constraint on the bandwidth required for data modulation can be relaxed, as the modulator can now operate at intermediate frequency (e.g. < 10 GHz). The technique furthermore yields a relatively high carrier-to-noise ratio (CNR) [Agr97]. On the other hand, equation (1.3) shows that the phase coherence between the optical signals is a critical issue for the implementation of an RHD scheme, as phase noises on the transmitter and LO lasers have a direct impact on the stability of the generated mm-wave range signal, even for intensity modulation schemes.

Due to the implementation of an LO laser at the receiver and, possibly, additional stabilization measures at the receiver, RAP design tends to gain in complexity and in cost, and a direct detection scheme as described in section 1.2.2.2 is often preferred [CIAHB97].

Remote heterodyning using a direct-detection receiver: In an effort to achieve both receiver simplicity and operability at mm-wave frequencies, heterodyning techniques can be combined with a DD receiver. In a simple heterodyning scenario, two optical modes from two lasers situated at the transmitter site are transmitted across the link and beat together on the photo-detector, where one of the two is modulated and the other serves as LO source. Depending on the target application, the stability of the two-laser setup may be sufficient even without further stabilization means, and bit rates up to 40 Gbps in IM links have been achieved using this approach [ZCY$^+$11]. A safer approach is usually that of *self-heterodyning*. By ensuring that the two modes beating on the detector originate from the *same* laser source, phase coherence can be improved, resulting in a more stable mm-wave signal.

Comb generators have been used to generate a broad spectrum from a laser source which is then filtered, leaving two comb lines separated by the desired mm-wave frequency beat together [SAPB10]. The comb generator, the optical filter and, where necessary, an optical amplifier must be added to the transmitter architecture.

Also, an MZM can be biased for frequency-doubling. Under this condition, the second harmonics of the seed frequency (e.g. 30 GHz) centered around the single-mode laser's optical wavelength beat together to give a signal at the desired mm-wave frequency (e.g. 60 GHz), and the central optical wavelength is suppressed ([NCLG09], [SPLG$^+$12]). The techniques is referred to as *optical double sideband with suppressed carrier* (ODSB-SC) transmission. Separate modulators need to be used for up-conversion and data modulation, and electrical signal generators are necessary. Therefore, transmitter architectures employing this technique tend to be complex.

All of the previous techniques share an increased hardware effort clearly to be avoided where RoF networks find large-scale application. With the ultimate goal of keeping systems costs to a minimum, a single optical device should perform mm-wave generation, ideally requiring simple bias circuitry.

A device which has the potential to satisfy these needs is the **mode-locked laser diode** (MLLD). This optical source exhibits a large comb spectrum of several tens of equidistant optical modes which are inherently locked in phase. If the separation of the optical modes is carefully defined during the design of the laser, a single mm-wave tone can be detected at an appropriate photo-detector when illuminated by the MLLD signal. In combination with the photo-detector, the MLLD thus represents an opto-electronic mm-wave oscillator providing the carrier signal for the RoF system.

The first error-free transmission experiments employing *actively*[18] mode-locked lasers in the 60 GHz range were performed as early as the mid-nineties, where sequences at 155 Mbps could be transmitted across 3 km of dispersion-shifted fiber ([BGvH$^+$96], [VHKKG97]). In the early 2000s, Hirata *et al.* demonstrated

---

[18]It will become apparent in section 2.1.2 that only passive or hybrid techniques are of interest regarding the target application.

the transmission of 3.0 Gbps using *hybridly* mode-locked lasers at a base frequency of 30 GHz in an analog optical link operating at a carrier frequency of 120 GHz [HHN03]. More recently, the first transmission experiments based on *passively* mode-locked quantum-dash lasers were carried out according to the IEEE 802.15.3c standard with a data rate of 3.03 Gbps ([LTC⁺09], [CCC⁺08]). Both direct modulation [HCC⁺08a] and external [HCC⁺08b] setups have been demonstrated, the striking advantage of the MLLD being its small size, combined with its up-conversion and modulation capabilities under DC bias conditions only. However, previous work has consisted in mere demonstrations of functioning setups.

Conclusion: Remote heterodyning is a useful technique for optical 60 GHz generation. Data modulation can then be performed at a much lower intermediate frequency within the bandwidth of state-of-the-art E/O converters. RHD can be used in combination with a direct detector if the phase stability of the beating signals is sufficient. The mode-locked laser diode provides a number of coherent optical modes and thus are suitable for use in an RHD RoF system. The MLLD is a promising miniature device which allows a considerable reduction of hardware effort in RoF transmitters. No detailed study on the feasibility of MLLD-based RoF networks has yet been conducted.

1.3 Problem statement and outline

The work presented herein is concerned with the study of analog optical fiber links operating in the 60 GHz range that employ a novel type of laser source: the quantum-dash mode-locked laser diode. This device is based on a large number of quantum dashes in the laser active medium and can provide electro-optical conversion and up-conversion to the 60 GHz in an extraordinarily simple configuration.

Quantum-dash mode-locked lasers exhibit unique properties that translate into system-level advantages over other types of sources, such as: high laser gain (i.e. high optical power), fast carrier dynamics (i.e. high modulation bandwidths), low threshold current (i.e. power-efficient operation) and narrow beat note linewidths (i.e. potential for low phase noise sources). Its particular interest for RoF transmitters is grounded in the following features:

- The MLLD provides mm-wave generation in the 60 GHz range under DC bias conditions.

- The MLLD exhibits a sufficiently large bandwidth for direct modulation.

- The MLLD is small in size and thus easy to integrate: the laser has a cavity length in the range of 700 µm for 60 GHz operation.

- No additional electrical mm-wave sources are needed. MLLD-based transmitters are thus inherently low cost.

While previous studies have shown the general possibility of employing such diodes in mm-wave RoF networks (references given in the previous section), a number of questions have remained open as to how to exploit the full potential of the devices for the target application. It is the purpose of this work to thoroughly investigate the performance of mm-wave RoF systems based on mode-locked laser diodes. The properties of mode-locked laser-based signal generation, modulation and transmission across analog optical links are investigated from a system designer's point of view in a feasibility study. Furthermore, it is the objective of this work to provide a set of guidelines for both component and system engineers working on quantum-dash MLLD suitable for employment in mm-wave RoF systems.

The feasibility study of 60 GHz RoF transmission systems based on quantum-dash mode-locked laser diodes encompasses several aspects which are treated in different chapters. At the beginning of each chapter, basic theoretical concepts and definitions are briefly reviewed. This is considered necessary as this work finds itself at the manifold interfaces of laser optics, optical communications and RF system engineering, and readers who are familiar with one of these domains may not be quite at ease with the others. The dissertation is organized as follows: In **Chapter 2**, the quantum-dash mode-locked laser diodes are presented. While for the scope of this work detailed knowledge of the laser's physics is not paramount, it will be necessary to study the general behavior of the laser diodes in order to understand their use at system level. The mm-wave generation process will be explained. The characterization results of the diodes in static and dynamic regimes will be presented. We will identify the principal limitations to their use in RoF systems, namely their frequency instability, and the RF power they can provide. The following chapters shall then be dedicated to a thorough discussion of the measures that can be employed to overcome those limitations.

Chapter 3 is dedicated to the analysis of link architectures and modulation techniques employing the MLLD. Direct and external modulation will be investigated in terms of parameters commonly used in RF system design, such as output power, link gain, system noise figure, signal-to-noise ratio and dynamic range. Techniques to enhance link performance will be presented.

Chapter 4 is dedicated to propagation effects. The degradation of the mm-wave signal due to attenuation and dispersion effects is studied with regard to the MLLD's fundamental properties, such as the width and shape of its spectrum, its phase noise and chirp characteristics. Furthermore, bandwidth-related constraints on propagation will be investigated in this chapter.

In **Chapter 5** deals with the problem of the inherent frequency instability of the passively mode-locked laser diode. A phase-locked loop based stabilization architecture is presented which provides a stable modulation sideband in the presence of high frequency jitter or phase noise on the carrier signal generated by the MLLD. The architecture presented herein can furthermore be employed as synthesizer for the sideband center frequency, a feature which comes in handy as the frequency selection of the MLLD in the manufacturing process is not sufficiently precise. The architecture is studied analytically, as well as through simulation and experiment. A system demonstrator has been developed; its function is validated both in the electrical domain and in the fiber-optic link.

Chapter 6 focuses on data transmission experiments in a MLLD-based RoF system. Error vector magnitude will be defined as a figure of merit for transmission quality. The impact of reduced power (i.e. low signal-to-noise ratio), and frequency jitter or phase noise on transmission quality are investigated within the context of the stabilized and the unstabilized RoF system. Transmission experiments are shown for data rates up to 4.234 Gbps.

In **Chapter 7**, the findings of this work are summarized and possible directions of future research are considered.

Notation and general remarks

Throughout the text, we deal with electromagnetic waves in a wide range of frequencies. **Optical** carriers result in beat signals in the **millimeter-wave** region. These beat signals provide the electrical carriers which in turn can be modulated in the **microwave** domain. The frequencies cover a range from a few GHz (10^9) to the hundreds of THz (10^{14}). Physicists and engineers working in the optical domain are used to describing the signals in terms of their mode wavelengths λ or optical frequency ν, while electrical engineers typically prefer the notion of frequency f. These distinctions are familiar yet might be confusing. For the scope of this work, we shall thus adhere to the following convention illustrated in figure 1.3: The variable ν will be used in the description of the mode spectrum of the optical signal. It will also serve to describe the frequencies of the beat signals ν_F when used in a direct context with the laser device providing the optical signal. Otherwise, f shall be used for the electrical frequency. The width of the spectral lines in the optical or electrical spectrum are $\Delta\nu$ for the optical linewidth and $\Delta\nu_F$ for the linewidth of the generated mm-wave carrier. The variable f shall furthermore be used in a more general way for all signals that are of interest in the electrical domain, in particular, in the description of the phase power spectral densities.

We furthermore deal with both DC (constant over time) and AC quantities (vari-

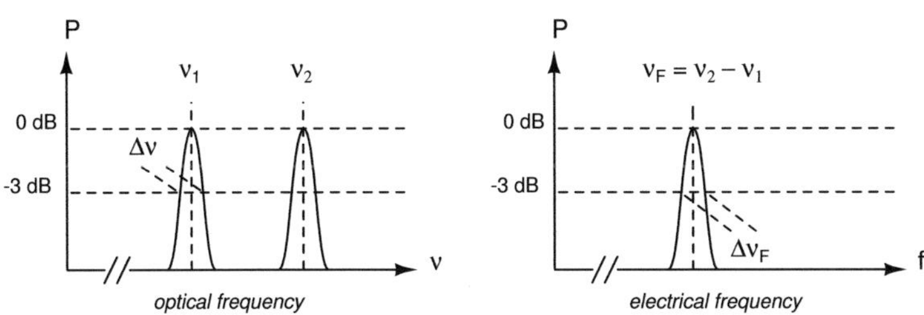

Figure 1.3: Notation

able with time). An upper case letter will be used for a DC quantity whereas a lower case letter will be used for an AC quantity (e.g. average optical power P_{opt} and RF power p_{RF}). In the context of a Fourier-transformed variable, an upper case letter denotes the description of a quantity in the frequency domain.

For circuits and diagrams, standard circuit symbols according to the IEEE Standard 315-1975 [IEE93] will be used at all times.

2 Mode-locked Laser Diodes for the Radio Over Fiber Link

Radio-over-fiber systems based on mode-locked lasers rely on the principle of remote heterodyning. The interference of two optical modes can be recovered as electrical beat signal by a simple direct-detection receiver. The key to designing and evaluating such RoF systems lies with the laser itself; all system-level considerations must therefore start out from an understanding of the operation of the mode-locked laser diode. While the theory of mode-locked lasers has been subject to extensive research for a number of decades and has been refined for a number of architectures, device physics - as typically manifested in rate equations - is not usually helpful for link design. This is why, in this chapter, the principle of operation of mode-locked lasers will be presented from a link designer's point of view. The laser is described with regard to its potential for fulfilling the task of a mm-wave oscillator and an E/O converter in the mm-wave RoF link. The interested reader might refer to [Hau00] or [KMS$^+$04] for details on ML laser physics. Link performance in terms of power or distortion is tightly linked to device performance. The electro-optical characterization of the diodes will be presented, followed by a discussion of a number of design parameters that must receive particular attention during the device design process for the specific purpose of the use in mm-wave RoF links. Finally, the limitations of present-day MLLDs, and how to overcome them, will be the focus of the last section of the chapter.

2.1 The mode-locked laser diode: Principle of operation

The mode-locked laser is a source of pulsed laser light and can be designed to emit ultrashort pulses in the range of femtoseconds under mode-locking conditions [KMS+04].

The term "mode-locking" refers to the phase coupling of the numerous modes of a *multi-mode[1]* laser spectrum. The phases of the different spectral components are said to be locked when they differ only by a constant phase shift at all times. The laser field can then be described as the Fourier-series extension of a periodic function of time, i.e. a pulse train. The separation in optical frequency v_F, known as the laser's *free spectral range* (FSR), corresponds to the pulse repetition frequency or the laser's *self-pulsation frequency[2]*.

Whether a time domain or a frequency domain approach might be more helpful depends on the target application. The time-based approach is of great use for applications that involve timing issues, e.g. optical sampling or clock generation. In these cases, short pulses and detailed knowledge about their dynamics are of primary interest. A thorough analysis on the topic can be found in [KMS+04].

In communication systems, mode-locked lasers have served as stable pulse sources in optical time division multiplex (OTDM) systems, or as multi-wavelength sources in wavelength division multiplex (WDM) networks. Other than that, mode-locked lasers have been used where high pulse intensities (micromachining and surface treatment) or very short pulses (imaging, electro-optical sampling) are needed [AMP00].

In this work, the mode-locked laser is employed (together with a suitable means of photo-detection) as mm-wave radio carrier generator. We shall therefore concentrate on the frequency domain characteristics of the device, such as its electrical beat spectrum and the RF phase noise displayed by the free-running laser, as they have a direct impact on the performance of the MLLD-based RoF system.

---

[1]Pulsed operation of a *single-mode* laser can be achieved based on the transient dynamics of the laser medium, such as the modulation of laser gain (gain-switching) or cavity losses (Q-switching) [ST07].

[2]The notions of free spectral range FSR, separation in optical frequency v_F, self-pulsation frequency and pulse repetition frequency essentially refer to the same property of the laser and are therefore used synonymously throughout the text.

2.1.1 The MLLD as non-ideal oscillator

A laser is first and foremost an optical oscillator. It comprises a resonant structure including an amplification section whose output is fed back to the resonator, whereby the phase is matched to fulfill the oscillation condition, i.e. the total round-trip phase shift is a multiple of 2π. Sustained oscillation is possible when the gain of the amplification compensates for the combined losses in the feedback system. As such, the laser oscillator does not conceptually differ from an electrical oscillator.

In a Fabry-Pérot (FP) laser, consisting of an active medium inserted between two mirrors, the cold resonator of length d sustains optical modes at frequencies

$$v_m = m \cdot \frac{c}{2d}, \tag{2.1}$$

where $m = 1, 2...$ and $c = c_0/n$ is the speed of light in the medium of refractive index n. For the scope of this work, the frequency $v_F = v_m - v_{m-1}$ lies in the mm-wave range. The supported longitudinal modes that will actually lase are determined by the frequency-dependent gain and loss curves of the resonator, effectively "windowing" the optical spectrum. They are centered around a frequency v_0 where the gain curve reaches its maximum.

Detecting the intensity of the FP multi-mode spectrum yields constant (though noisy) electrical power due to the random nature of the phase relation between the oscillating FP modes.

If, however, the optical modes can be made to oscillate in a coherent manner, i.e. "mode-locked", the detected intensity will yield a time-variant electrical signal with spectral components at the beat frequency v_F and its harmonics.

Any real (i.e. non-ideal) electrical or optical oscillator suffers from noise effects spreading the lineshape of the otherwise well localized resonances. The phenomenon of oscillator line broadening has been intensively studied, and a number of parameters such as the gain and loss characteristics, as well as spontaneous emissions or the emission wavelength, play a role in determining its actual value[3] ([ST58], [Hen82]). In the case of the mode-locked multi-mode spectrum, both the optical modes and the spectral lines of the electrical beat signals therefore each exhibit finite linewidth. The notion of linewidth in this context refers to the full width at half maximum (FWHM) of the spectral line. For this reason, it is also referred to as 3 dB-linewidth.

2.1.1.1 Pulsed operation

For the following consideration, it is helpful to describe the laser spectrum as symmetrically centered around a center frequency v_0, counting the modes from the center using negative and positive indices. This is also the preferred notation used in the references cited hereafter though we will later abandon this description for practical reasons.

For a total of M modes around v_0, the amplitude of the complex waveform propagating in z direction in the cavity can be expressed as [Cha11]

$$e_{\mathrm{opt}}(z,t) = \sum_{|m| \leq \frac{M}{2}} |A_{\mathrm{m}}| \exp\left(j2\pi v_{\mathrm{m}}t - j\varphi_{\mathrm{m}}\right) \exp\left(-k_{\mathrm{m}}z\right). \tag{2.2}$$

Here, $|A_{\mathrm{m}}|$ and φ_{m} represent the amplitude and the phase of a mode m, v_{m} and k_{m} its frequency and its wave vector, respectively, where $v_{\mathrm{m}} = v_0 \pm mv_{\mathrm{F}}$.

The center frequency v_0 is in the order of about 194 THz at a wavelength of 1.55 µm, while the width of the spectrum Mv_{F} is in the order of 2 THz (i.e 30 - 40 modes separated by about 60 GHz). Thus, Mv_{F} is a small variation around v_0, and the wave vector k_0 can be linearized by means of a Taylor series expansion,

$$k_{\mathrm{m}} = k(v_0) + \left(\frac{\partial k}{\partial v}\right)_{v_0} (v_{\mathrm{m}} - v_0) = k_0 + \frac{2\pi m v_{\mathrm{F}}}{v_{\mathrm{g}}}, \tag{2.3}$$

where v_{g} is the group velocity of the wave at the center frequency v_0. Introducing the phase velocity $v_{\varphi} = v_0/k_0$, the waveform from (2.2) becomes

$$e_{\mathrm{opt}}(z,t) = e_{\mathrm{RF}}(z,t) \cdot \exp\left(j2\pi v_0 \left(t - \frac{z}{v_{\varphi}}\right)\right), \tag{2.4}$$

---

[3]In the case of the single-mode laser, the Schawlow-Tones lineshape represents a fundamental limit [ST58].

where $e_{RF}(z,t)$ is the complex envelope, or the RF field, modulating the optical wave at v_0. It is given by

$$e_{RF}(z,t) = \sum_{|m| \leq \frac{M}{2}} |A_m| \exp\left(jm2\pi v_F \left(t - \frac{z}{v_g} \right) \right) \exp(j\varphi_m). \qquad (2.5)$$

The amplitudes $|A_m|$ can be determined from the spectral profile of the gain and the cavity losses. In the case of a multi-mode laser, such as e.g. a simple FP cavity, the phases φ_m are random in time and each mode propagates independently from its neighbors. If however the φ_m are locked, i.e. the phase differences between adjacent modes $\Delta\varphi_{m,m-1}$ are constant, and if the $|A_m|$ are properly chosen, $e_{RF}(z,t)$ will take on the form of periodic pulses.

For a perfectly flat laser spectrum of M modes ($m = 0, \pm 1, \pm 2 ... \pm k$, and $M = 2k+1$, and $|A_m| = A$), it can be easily shown by means of a geometric series that the envelope takes on the form [ST07]

$$e_{RF}(z,t) = A \frac{\sin\left(M\pi v_F \cdot \left(t - \frac{z}{v_g} \right) \right)}{\sin\left(\pi v_F \cdot \left(t - \frac{z}{v_g} \right) \right)}. \qquad (2.6)$$

The optical intensity I associated with the waveform is then

$$I(z,t) \propto |e_{RF}(z,t)|^2 = A^2 \frac{\sin^2\left(M\pi v_F \cdot \left(t - \frac{z}{v_g} \right) \right)}{\sin^2\left(\pi v_F \cdot \left(t - \frac{z}{v_g} \right) \right)}. \qquad (2.7)$$

$I(z,t)$ is a periodic function, i.e. a pulse train of the repetition frequency v_F mentioned earlier, as illustrated in figure 2.1. Its peak intensity is M times the average intensity $\bar{I}$, and the pulse duration at half width is $\tau = 1/(v_F M)$.

When the MLLD is used in a remote heterodyning concept, it is more intuitive to describe the laser in the frequency domain. Modulation and mixing of the laser modes can also be illustrated effectively in the frequency domain. The Fourier transform of the complex waveform is the convolution of the Fourier transform of the complex envelope and the remaining optical field. It will often be sufficient to only consider the contribution of the RF envelope, $\mathscr{F}\{e_{RF}(z,t)\} = E_{RF}(v)$,

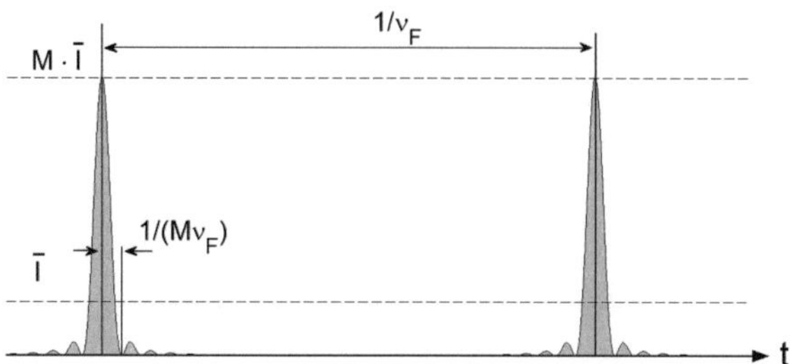

Figure 2.1: MLLD pulse train

and otherwise regard the modes as perfectly harmonic oscillations. Nevertheless, one should keep in mind throughout further discussions that the mode-locked spectrum is indeed a comb spectrum with components of **finite** linewidth centered around the optical center frequency ν_0. The spectral characteristics of the MLLD shall be the topic of the next section. Before regarding the MLLD, it is worth highlighting some general properties of the (optical or electrical) oscillator spectrum, as such an analysis will also be found useful in later chapters.

2.1.1.2 General remarks on the oscillator spectrum in the presence of noise

The spectrum of an oscillator is perturbed by two basic types of noise: *Additive* noise such as thermal noise in electrical components, is always present; it is in general white. Likewise, white noise originating from spontaneous emission cannot be avoided in laser oscillators. *Parametric* noise requires the presence of a carrier and consists in a near-DC process modulating the carrier in amplitude, or in phase, or in both. Among the different kinds of parametric noise, the *flicker* or f^{-1} noise process is dominant. Other types of parametric noise include noise originating in power supplies or temperature fluctuation [Rub09].

The determinant noise form with regard to the oscillator spectrum is phase noise. It is commonly described by the power spectral density (psd) $S_\varphi(f)$ of the random phase fluctuation $\varphi(t)$. It is equally possible - and often done - to describe oscillator noise in terms of the fractional frequency fluctuation $y(t) = \Delta v(t)/v_0$ relative to a carrier v_0. The corresponding psd then becomes

$$S_y(f) = \frac{f^2}{v_0^2} S_\varphi(f). \tag{2.8}$$

In analog electronics, the most widely used model of the oscillator phase noise is arguably that of Leeson[4] which describes the phase noise psd by means of a sum of power-law processes

$$S_{\varphi,\zeta}(f) = \sum_{\zeta=0}^{4} k_\zeta \cdot f^{-\zeta}, \tag{2.9}$$

yielding the well-known straight-line tangential approximations on a log-log plot [Lee66]. It conveniently illustrates the impact of different sources of noise which are predominant in certain regions at an offset relative to the oscillation frequency:

- $\zeta = 0$: white phase noise; with an f^0 dependency in the phase psd

- $\zeta = 1$: flicker phase noise; with an f^{-1} dependency in the phase psd

- $\zeta = 2$: white frequency modulated phase noise; with an f^0 dependency in the frequency psd and an f^{-2} dependency in the phase psd (also referred to as random walk phase modulated phase noise)

- $\zeta = 3$: flicker frequency modulated phase noise; with an f^{-1} dependency in the frequency psd and an f^{-3} dependency in the phase psd

- $\zeta = 4$: random-walk frequency modulated phase noise; with an f^{-2} dependency in the frequency psd and an f^{-4} dependency in the phase psd

---

[4]While this heuristic approach is of considerable practical value, it remains unsatisfactory in explaining the underlying phase noise generation mechanism, as it implies infinite noise power at zero frequency. Furthermore, the superposition of the influences of multiple noise sources to give the oscillator's over-all phase response has been proven to be inaccurate for low-frequency perturbations close to the carrier [DMR00]. The contributions of the various noise sources on the phase noise psd are correctly considered through convolution rather than summation, although the summation provides a good approximation for small-angle phase perturbations [CB06].

In Leeson's model, the quality factor Q of the oscillator is a useful parameter for understanding the operation of an oscillator. The quality factor of an oscillator is given by the resonator frequency f_r and the resonator bandwidth Δf_r:

$$Q = \frac{f_r}{\Delta f_r}. \tag{2.10}$$

For a laser, f_r is determined by the material system. Δf_r is essentially determined by the photon lifetime (and in turn, by the transmission losses of the mirrors in the cavity); for weakly damped oscillations, i.e. high-Q lasers, it is defined as the FWHM (-3 dB) bandwidth of its resonances[5].

For an electrical oscillator, f_r is determined by the capacitance and the inductance constituting the resonator, while Δf_r depends again on the resonator losses, and is again defined at the - 3 dB cut-off of the response.

Among the different noise processes, the ones indexed $\zeta = 0, 1, 2$ are paramount for the oscillator psd:

White phase noise, $\zeta = 0$: Far from the carrier, the psd $S_{\varphi, \zeta=0}$ can be expressed in terms of the oscillator's noise figure F, at a reference temperature of $T_0 = 290$ K

$$S_{\varphi, \zeta=0}(f) = \frac{k_B T_0 F}{P_C}, \tag{2.11}$$

where P_C is the carrier power, k_B is Boltzmann's constant and $k_B T_0$ takes on the value of $4 \cdot 10^{-21}$ J, or -174 dBm/Hz. At large offset frequencies from the carrier, the phase noise spectrum is thus white, due to background thermal noise and the active-device transducer gain (spontaneous emission of light in a laser medium, or thermal motion of electrons in an electrical circuit).

Flicker phase noise, $\zeta = 1$: Numerous physical fluctuations have empirical (measured) spectral densities proportional to f^{-1}. At large offsets from the car-

---

[5]Often, Δf_r is quoted as the ratio of the *finesse* of the laser, and its free spectral range. The FHWM bandwidth definition re-enters the equation via the definition of the finesse itself: it corresponds to the free spectral range divided by the FWHM bandwidth of its resonances [Pas08].

rier, this phase noise disappears in the flat white noise. Flicker phase noise is the signature of non-oscillatory components such as amplifiers or frequency dividers. Although not yet fully understood, it is generally assumed that the phase flicker characteristic in oscillators stems from near-DC f^{-1} noise, up-converted to the carrier region [RB10]. Phase fluctuations have been modeled by a Gaussian random walk[6] whose mean-square deviation increases without a bound. As such, the f^{-1} phase noise has been approximated by a Lorentzian[7] function, see below [Dem02]. This approach has also been adopted in [Her98], where the analogy between a laser and a voltage-controlled oscillator was shown.

White FM phase noise, $\zeta = 2$: The case of white frequency modulated (FM) phase noise is well-studied [TSG$^+$04]. It is commonly referred to as *timing jitter* in time-domain representations. A CW oscillation suffering white FM phase noise takes on the shape of a Lorentzian. In terms of the oscillator quality factor, the corresponding psd is

$$S_\varphi(f) \sim \frac{1}{(1 - r_{\mathrm{G}})^2 + \left(\frac{2Q}{f_{\mathrm{r}}}\right)^2 (f - f_{\mathrm{r}})^2}. \tag{2.12}$$

Here, $r_{\mathrm{G}} \leq 1$ represents the relative gain, the ratio of resonator gain and total loss. Following the notation of [Her98], the linewidth of the Lorentzian is

$$\Delta f = (1 - r_{\mathrm{G}}) \frac{2Q}{f_{\mathrm{r}}}; \tag{2.13}$$

it is often determined experimentally. Equation (2.12) can be stated as

$$\widetilde{S}_\varphi(f) \sim \frac{\Delta f}{(\Delta f)^2 + (f - f_{\mathrm{r}})^2} \tag{2.14}$$

which corresponds to the notation commonly found in the discussion of mode-locked lasers [KMS$^+$04].

---

[6]The random walk is a mathematical formalization of a trajectory that consists of taking successive random steps; a Gaussian random walk relies on step sizes that vary according to a normal distribution.

[7]A Lorentzian is the shape of the squared magnitude of a one-pole low pass filter function.

Note that flicker FM phase noise ($\zeta = 3$) and random-walk FM phase noise ($\zeta = 4$) are long-correlated events in the oscillator phase. They will not be considered here, as low-frequency perturbations are of secondary interest in the scope of this work. It should however be stated that both will generate Gaussian-like components in the oscillator psd, leading to a lineshape considerably wider than the Lorentzian close to the carrier. The interested reader is referred to [Her98] and [CB06].

The mode-locked laser is an optical oscillator providing not only one oscillation at one discrete frequency, but several locked modes at equally spaced frequencies. As a consequence, the output intensity is pulsed. In combination with a band-limited photodiode detecting a certain component of the electrical beat spectrum, it can be employed as an electrical oscillator whose spectrum can be described in terms of the above-mentioned regions. The concept of the Lorentzian lineshapes is quite useful in the understanding of the mode-locked laser spectrum and its beat spectrum.

For practical concerns, such as in chapter 5, we will employ the tried and tested Leeson model containing f^0, f^{-1}, f^{-2}, f^{-3}-dependencies.

2.1.1.3 Optical spectrum and beat spectrum

The shape of the optical spectrum of a passively mode-locked semiconductor laser is mainly influenced by the gain and loss characteristics of the cavity, which effectively select the resonator modes that will actually lase and form the comb spectrum.

In a real laser, noise will influence the amplitude of the lasing modes, the center frequency v_0, the separation between the modes of the comb v_F, and the optical phases of the modes. Timing jitter (or white frequency noise) is a property of particular importance in the passively mode-locked laser, where mode-locking is achieved without the help of an external modulation that could force a pulse into a fixed timeslot [HOR$^+$09].

Takushima *et al.* have studied the phase noise profile of the fundamental note in the electrical beat spectrum. According to their findings, the profile is the result of three different processes which each exert a direct influence on line broadening as displayed in figure 2.2: (a) the conversion of the timing jitter of the optical pulses into electrical frequency noise, (b) the conversion from amplitude noise on the pulse train to frequency noise on the modes (AM/FM conversion) due to fast changes in carrier density and refractive index, and (c) a direct contribution of spontaneous laser emission, i.e. optical phase noise ([TSG$^+$04], [KOT$^+$08]). The broadening of the comb lines, described by the linewidth Δv_{m}, is dominated by the contributions of optical phase noise ($\zeta = 1$ noise process) and timing jitter ($\zeta = 2$ noise process). The resulting optical spectrum consists of M modes with Lorentzian line shapes:

$$S_{\mathrm{opt}}(v) \sim |E_{\mathrm{RF}}(v - v_0)|^2 \sum_{|m| \leq \frac{M}{2}} \frac{2\Delta v_{\mathrm{m}}}{(v - v_{\mathrm{m}})^2 + \Delta v_{\mathrm{m}}^2}, \qquad (2.15)$$

where $E_{\mathrm{RF}}(v)$ is the Fourier transform of the RF envelope of the laser pulse, and the linewidths, defined as the FWHM linewidth, vary with the mode number [TSG$^+$04]. At the photo-detector, the intensity of the optical pulse is detected, and the resulting spectrum at mm-wave frequencies is [HOR$^+$09]

$$S_{\mathrm{RF}}(v) \sim \sum_{|m| \leq \frac{M}{2}} \frac{2\Delta v_{\mathrm{F,m}}}{(v - m v_{\mathrm{F}})^2 + \Delta v_{\mathrm{F,m}}^2}. \qquad (2.16)$$

The Lorentzian shape is found as well in the mm-wave spectrum and there is a direct relationship between the optical linewidths of the modes, and the linewidth $\Delta v_{\mathrm{F,1}}$ of the fundamental beat note at v_{F} in the electrical spectrum, which is of course the one of interest:

$$\Delta v_{\mathrm{F,1}} = \frac{\Delta v_{\mathrm{m}} - \Delta v_0}{(m - m_0)^2}, \qquad (2.17)$$

where m_0 is the mode number corresponding to the mode with minimum linewidth Δv_0.

The optical and electrical linewidths, Δv and $\Delta v_{\mathrm{F,1}}$, are small and in general difficult to observe as they are broadened by other forms of noise, such as thermal

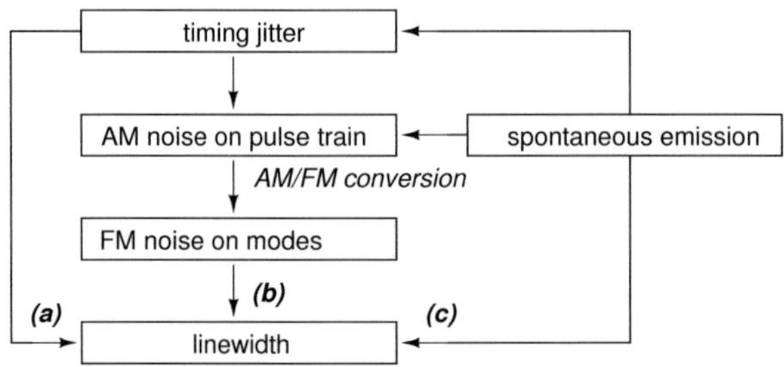

Figure 2.2: Contribution to linewidth broadening and electrical phase noise, from [TSG$^+$04]

drift, noise in the pump source and the like. Appropriate feedback measures or stabilization architectures must usually be employed (see chapter 5). On the other hand, linewidth is a concept well established in laser device characterization, but it is not in general a common property to observe in oscillator design, still less in system design. What is of interest here is usually the single sideband phase psd $L(f)$ of the generated mm-wave signal at a center frequency ν_F which will be the topic of section 2.2.3.2.

2.1.1.4 Millimeter-wave generation through heterodyning at the photo-detector

We abandon the description of the mode-locked laser spectrum as symmetrically centered around ν_0, and choose to rearrange the mode numbers. Mode 1 is the leftmost mode, mode 2 follows, and their beating together results in the leftmost mode of the electrical beat spectrum. The construction of the beat spectrum can then easily be implemented algorithmically, if we subsequently apply this procedure to all pairs of modes. Equation (2.2) is stated again with rearranged mode numbers, now running from $m = 1, 2...M$. Note that we have dropped explicit statement of the time variable t and the wave location z:

$$e_{\text{opt}} = \sum_{m=1}^{M} |A_{\text{m}}| \exp\left(j2\pi\nu_0 t\right) \exp\left(j2\pi \cdot m\nu_{\text{F}} t\right) \exp\left(j\varphi_{\text{m}}\right). \qquad (2.18)$$

Electrical mm-waves are generated when directly detecting the intensity, i.e. the RF envelope of the MLLD pulse by means of a photo-detector (PD) with a responsivity of ρ. The laser pulse carries the interferences of all laser modes.

The first question to resolve is that of the ideal optical spectrum yielding maximum power transfer to the electrical domain by means of a direct detection process. When the detector receives an optical power[8] p_{opt}, a photocurrent is generated:

$$i_{\text{ph}} = \rho\, p_{\text{opt}}. \qquad (2.19)$$

p_{opt} is the optical power on the PD, with an effective detection area A, and I the intensity in W/m^2 of the incoming plane lightwave, it is

$$p_{\text{opt}} = I \cdot A = \frac{1}{2} c \cdot \varepsilon_0 \left| e_{\text{opt}} \right|^2 \cdot A. \qquad (2.20)$$

According to equation (1.3), the squared magnitude of the electrical field is

$$e_{\text{opt}} \cdot e_{\text{opt}}^* = \left| e_{\text{opt}} \right|^2 = \sum_{m=1}^{M} |A_{\text{m}}|^2 + 2 \sum_{m=2}^{M} \sum_{n=1}^{m-1} |A_{\text{m}}| |A_{\text{n}}| \cos\left(2\pi(m-n)\nu_{\text{F}} t + \Delta\varphi_{\text{m,n}}\right).$$

$$(2.21)$$

The total generated electrical current contains a DC and an RF contribution, such that

$$p_{\text{el,tot}} = P_{\text{DC}} + p_{\text{RF}}. \qquad (2.22)$$

The photocurrent through the load $i_{\text{ph}} = I_{\text{ph,DC}} + i_{\text{ph,RF}}$ also contains a DC and an RF contribution. The DC contribution,

$$I_{\text{ph,DC}} = \frac{1}{2}\rho c \cdot \varepsilon_0 \cdot A \cdot \sum_{m=1}^{M} |A_{\text{m}}|^2, \qquad (2.23)$$

---

[8]Optical power in a classical interpretation refers to the average power obtained by taking the average of the instantaneous power over a few optical periods [Fre09].

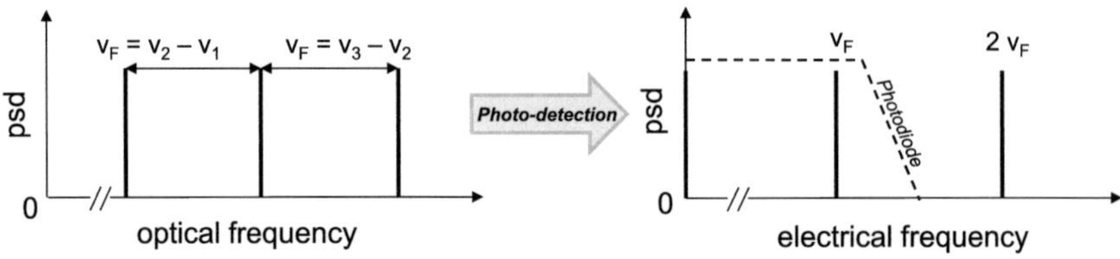

Figure 2.3: Principle of direct band-limited photo-detection

is of great importance when determining the detector's noise level. The topic will
be covered in chapter 3. At this time, we will focus only on the RF contribution in
the photocurrent. The electrical RF current flowing through the load resistance
R_0 matched to the PD exhibits spectral components at the k harmonics of v_F
($k = 1, 2, ...M − 1$):

$$
\begin{aligned}
i_{\text{ph,RF}} &= \frac{1}{2}\rho c \cdot \varepsilon_0 \cdot A \left(\sum_{m=2}^{M} \sum_{n=1}^{m-1} |A_{\text{m}}| |A_{\text{n}}| \cos\left(2\pi(m-n)v_F t + \Delta\varphi_{\text{m,n}}\right) \right) \\
&= \sum_{k=1}^{M-1} I_k \cos\left(2\pi v_F \cdot kt + \phi_k\right).
\end{aligned}
\tag{2.24}
$$

When the PD is band-limited with a bandwidth $v_F < BW_{\text{PD}} < 2 \cdot v_F$, the beat
signal of two neighboring modes is recovered at the fundamental beat note v_F,
providing the mm-wave carrier signal for the RoF system. The principle of band-
limited detection is shown in figure 2.3. Beat notes at frequencies $k \cdot v_F$, $k \geq 2$ are
filtered by the PD so that only a fraction $p_{\text{RF,c}}$ of the total generated RF power p_{RF}
will be recovered. When the MLLD serves as a photonic source for remote mm-
wave delivery, this filter phenomenon is highly desired. However, the electrical
power carried by the higher harmonics represents, like the generated DC power,
a net power loss in the generation process. We shall refer to this circumstance as
the *low-pass filter effect* induced by the band-limited PD.

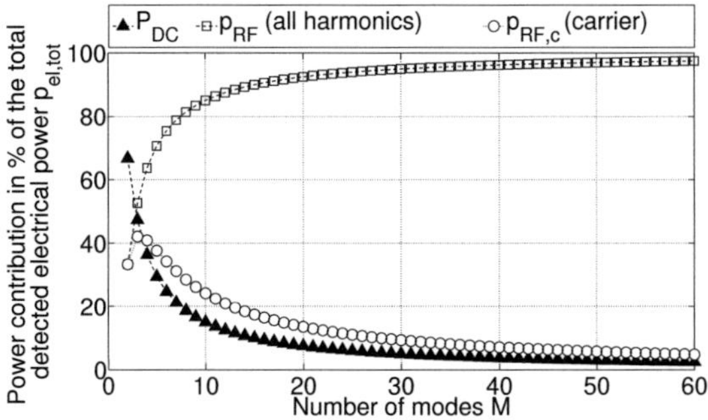

Figure 2.4: DC (triangles) and RF (squares and circles) contributions to the total electrical power $P_{\text{el,tot}}$ recovered by the PD. Previously published by the author in [BYP$^+$12].

The part of the optical field detected in the band-limited case is

$$\left|e_{\text{opt}}\right|^2 = \sum_{m=1}^{M} |A_{\text{m}}|^2 + 2 \sum_{m=1}^{M-1} |A_{\text{m}}|\,|A_{\text{m}+1}|\cos\left(2\pi \nu_{\text{F}}t + \Delta\varphi_{\text{m,m}+1}\right). \tag{2.25}$$

The carrier generated at the fundamental beat note appears in the current,

$$i_{\text{ph,RF,c}} = I_1 \cos\left(2\pi \nu_{\text{F}}t + \phi_1\right). \tag{2.26}$$

Assuming $\Delta\varphi_{\text{m,m}+1} = 0$, a comparison of the coefficients gives $\phi_1 = 0$ and

$$I_1 = \frac{1}{2}\rho \cdot c \cdot \varepsilon_0 \cdot A \sum_{m=1}^{M-1} |A_{\text{m}}|\,|A_{\text{m}+1}|. \tag{2.27}$$

The carrier has a power of

$$p_{\text{RF,c}} = R_0 \left(\frac{I_1}{\sqrt{2}}\right)^2. \tag{2.28}$$

It can easily be shown that the number of modes M must be reduced in order to limit this inherent loss. Fig. 2.4 shows the contribution of DC (P_{DC}) and

RF (p_{RF}) power to the over-all detected electrical power $p_{el,tot}$ after the PD, as well as the percentage recovered at the desired mm-wave frequency ($p_{RF,c}$) for variable M, $|A_m|$ = constant, and $\Delta\varphi_{m,m+1} = 0$. Supposing equal optical power p_{opt}, maximum relative carrier power is obtained for $M = 3$. As will be discussed in chapter 3, the same low-pass effect will limit the power available in the IM sidebands.

2.1.2 Active and passive mode-locking

So far, the question as to how the phase-locking of the optical modes can be achieved has remained open. In general, mode-locking can be attained by modulating one of the laser parameters at the pulse repetition frequency ν_F (active mode-locking). To this end, an active modulator is often included in the cavity. In lithium niobate modulators, the electro-optical effect[9] can be exploited, and the propagating pulse is reshaped once per round-trip triggered by an external modulating signal [AMP00]. Active mode-locking can only be used to stabilize the laser at its own self-pulsation frequency but offers very limited tunability.

Mode-locking can also be achieved by exploiting the properties of a non-linear medium inserted into the cavity, e.g. a saturable absorber (passive mode-locking). The absorber is a medium whose absorption coefficient decreases, as the intensity of the light passing through it increases [ST07]. It will thus block a weak signal, whereas a relatively powerful pulse will be reshaped repetitively when it circulates through the cavity. Yet another passive mode-locking technique relying on non-linear optics makes use of the Kerr effect, where the intensity of the passing light modifies the refractive index of a Kerr lens and thus, an artificial saturable absorber section is created [NSGF91].

Hybrid mode-locking techniques exist where both techniques are combined: Passive mode-locking is then enhanced by applying an external modulation at the laser's self-pulsation frequency. The phase noise of the resulting beat note can so be improved considerably [FAB+10].

Passive mode-locking without a dedicated absorber or lens section has been demonstrated in [RBD+05] and [LDR+07], where the self-oscillation of the

---

[9]For details, see appendix B.

laser diodes is attributed to four-wave mixing in an active medium consisting of stacked layers of quantum dots. This type of passive mode-locking is of particular importance to this work, as the MLLDs herein are based on the same technology.

It is obvious that in the event where the MLLD is supposed to replace an electrical mm-wave oscillator, a technique incorporating active mode-locking is of little interest. The technique might however be of service when stable lock can be achieved by applying only a *sub-harmonic* of the desired pulse rep tition frequency.

2.1.3 Quantum-dash MLLD: Technology and architecture

The effects of 3D carrier confinement in the active zone of a laser were first studied by Arakawa and Sakaki [AS82]. They predicted the creation of quantum-dot devices exhibiting significantly lower threshold current densities, as well as a high independence of laser operation from temperature variations. Further advantageous properties include fast carrier dynamics - translating into enhanced direct modulation capability at system level - and high modal gain. For about three decades, considerable efforts have been devoted to the experimental realization of semiconductor heterostructures that enable 3D carrier confinement. In many aspects, quantum dots yield an enormous potential and represent a great challenge in semiconductor physics. Their properties resemble those of atoms locked "in an electromagnetic cage", as Bimberg *et al.* phrased it in their comprehensive study of quantum dot heterostructures [BGL98]. Elongated dots referred to as *quantum dashes* provide the physical basis for the mode-locked laser devices that were used in the course of this thesis project. They certainly represent a worthy field of study in their own right. The reader is referred to the relevant literature for details [RCS07], [BGL98].

The quantum dot laser was patented in 1993, consisting - in its most basic configuration - of a plurality of quantum dots disposed in a laser host material, and a pumping source for exciting and inducing a population inversion in the quantum dots [HBTH]. Passive mode-locking in quantum-dash lasers was first demonstrated in 2001, with pulses of approximately 17-ps duration at a wavelength of 1.3 μm and a repetition rate of 7.4 GHz [RCS07].

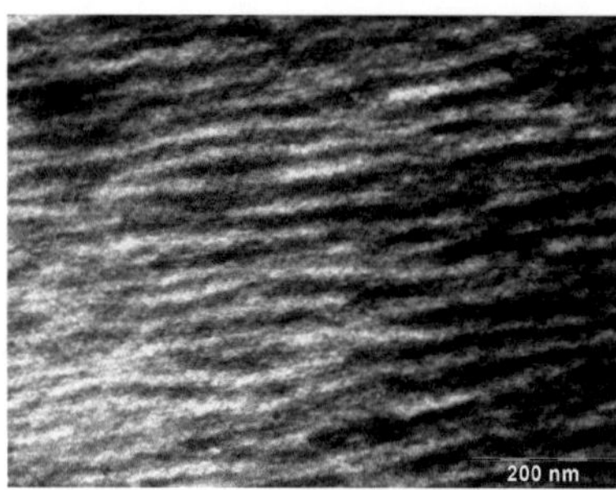

(a) InAs/InGaAsP quantum dashes

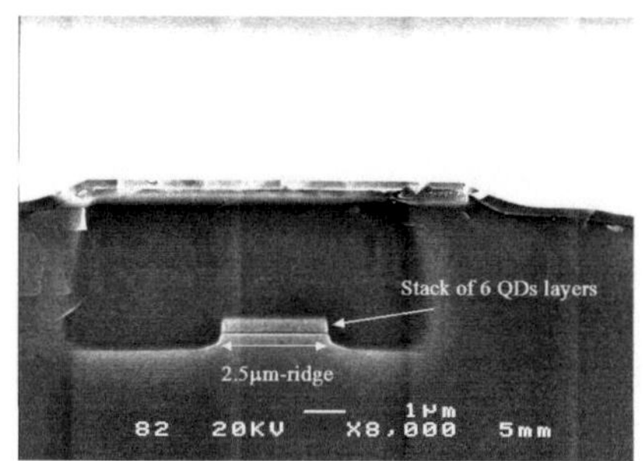

(b) Fabry-Pérot laser architecture, dash-in-a-barrier.

Figure 2.5: Transmission electron microscopy (TEM) images of InAs/InGaAsP quantum dashes, Fabry-Pérot architecture: Courtesy of G. Patriarche, CNRS LPN.

Today, component designers still face technological challenges such as the question of how to obtain high modal gain which, in turn, implies low threshold current densities. The interaction between the optical modes and the quantum dashes, resulting in high gain, can be improved by stacking several layers containing quantum dashes. Another challenge is to obtain temperature insensitivity, often achieved (among other solutions) through p-doping [MB06][10]. For a final challenge, minimizing the beating linewidth is a crucial issue of mode-locked laser design. The beating linewidth is owed to the small confinement factor (i.e. to the small geometrical ratio of the active region volume to the modal volume) rather than to the 3D quantification of electronic energy levels. Details on technology and fabrication can be found in [LDR+07].

The quantum-dash heterostructures used in this work were grown on S-doped InP wafers using a self-organized[11] growth mode. Its principle is based on strain relaxation occurring in a very thin (≤ 1 nm) highly strained InAs layer deposited

---

[10] A common p-dopant is sulfur (represented by S in the periodic table).

[11] A method referred to as the Stransky-Krastanow growth, see [BGL98].

on an InGaAsP layer. The mismatch between the material lattice constants of typically 4% in this material system induces the formation of quantum dashes. Figure 2.5(a) shows a transmission electron microscopy (TEM) image of such a structure. The dimension along the growth axis, controlled by the nominal thickness of the InAs layer to achieve 1.55 µm emission, is about 2 nm. The typical width of the dashes is about 15 to 20 nm. The length of the dashes ranges between 40 and 300 nm. The active zone of the MLLD consists of six stacked layers of quantum-dashes, each of a height of approximately 10 nm. Figure 2.5(b) shows a TEM image of the laser. Stable mode-locking is observed under DC supply conditions in a single-section Fabry-Pérot cavity.

2.2 Characterization

The MLLD were manufactured by III-V Lab, Joint Laboratory of Alcatel Lucent Bell Labs, Thales Research & Technology in Palaiseau, France, and delivered as unpackaged chip diodes. The three samples that have been used for this work are labeled by the identifiers L34, L611 and L872. Each chip is supplied on a rectangular submount of a size of 2 mm x 6 mm and connected by means of a single wirebond to a coplanar waveguide (CPW) transmission line on an aluminum nitride (AlN) substrate. As the chips originate from the same manufacturing process, they exhibit similar properties which is why we consider it appropriate to show characteristic results that have been obtained on one sample as representative for all three samples.

2.2.1 Laser testbench

In order to characterize the diodes, an appropriate testbench was conceived. A picture of the testbench is shown in figures 2.6 and 2.7 where two different coupling systems, namely a micro-lense SMF and a focaliser, respectively, have been used. The elements of the testbench will be explained further in section 2.2.1.2.

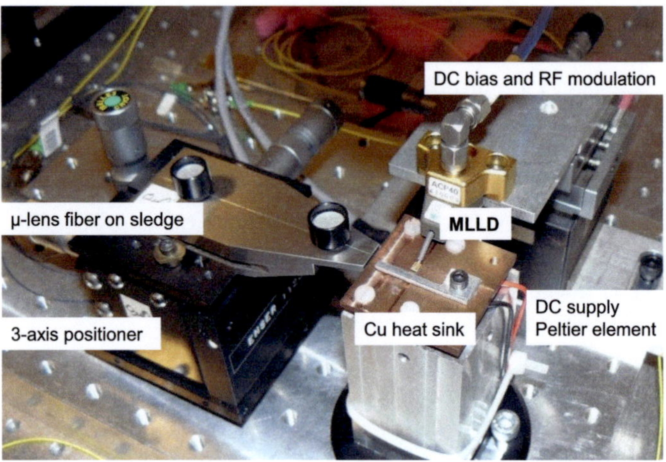

Figure 2.6: Testbench for MLLD with micro-lense fiber and coplanar GSG probe

The RF connection is established by probing directly onto the CPW line[12] (see in particular, figure 2.7). The submount is then placed on a copper (Cu) groundplate which serves as a heat sink. The probe is connected to the pump current supply through the DC input of a bias-T. The modulation signal for operation in dynamic regime is supplied via the RF input of the bias-T.

2.2.1.1 Temperature control

Despite early theoretical predictions, the operating characteristics of the MLLD, such as emission wavelength and threshold current, but also mode-locking itself, vary with temperature, and the stable operation of the diodes requires that temperature be regulated. Controlling the temperature of the devices also makes it possible to evaluate such heating or cooling effects. Solid-state thermoelec-

---

[12]Cascade Microtech Infinity Probe GSG 150

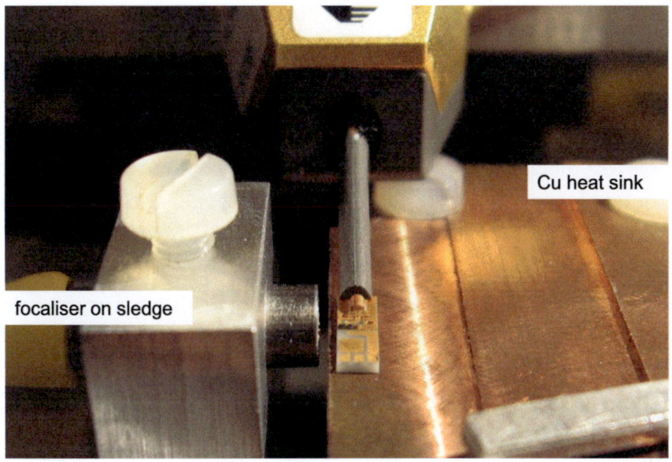

Figure 2.7: Testbench for MLLD with dual lens focaliser and coplanar GSG probe

tric devices based on the so-called *Peltier* effect[13] are used in active electronic feedback loops: the actual temperature is measured by a sensor, here, a negative-temperature-coefficient (NTC) thermistor. The temperature is compared to a set-point temperature, and an error signal proportional to the difference is produced. The error signal is fed into an appropriate bipolar driver which is connected to the thermoelectric device. In the testbench, the thermoelectric device is placed directly below the Cu plate, and the sensor is inserted into a small hole inside the plate. The operating temperature is adjusted to 25° Celsius.

2.2.1.2 Chip-to-fiber coupling

Due to its diffraction at the emitting edge of the chip, the laser light is radiated into a comparatively wide volume. At the edge, the waveguide cross section corresponds to a rectangular aperture, causing the light to diverge in both vertical and horizontal planes, and resulting in an elliptical beam cross section. The wide divergence can be problematic if a fiber with a small numerical aperture and a small core radius is used, such as SMF, because a large portion of the light will be radiated into free space rather than coupled into the fiber.

The optical coupling between the laser chip and the fiber is characterized by the transmission T and the reflection R. Their logarithmic values are referred to as *insertion loss* and *return loss*, respectively. Simple or multiple optical reflections due to poor coupling can severely impact the lasing process in the cavity and might disturb mode-locking. A direct effect is an increase in transmitter (phase) noise and, in consequence, a degradation of signal quality at the receiver. For analog transmission systems using intensity modulation, a return loss of about 55 dB or more is usually required for single-mode sources [Fis02]. It is very probable that the return loss must be even higher for the highly sensitive mode-locked lasers, but it appears that detailed studies have not yet been carried out on the topic.

Reflection can be mitigated by the use of an optical isolator. All characterization and system measurements shown in here include an optical isolator which is con-sidered a part of the transmitter module, and which introduces an insertion loss

---

[13]Depending on the direction of the electrical current passing through a junction of dissimilar metals, heat can be created or absorbed at the junction.

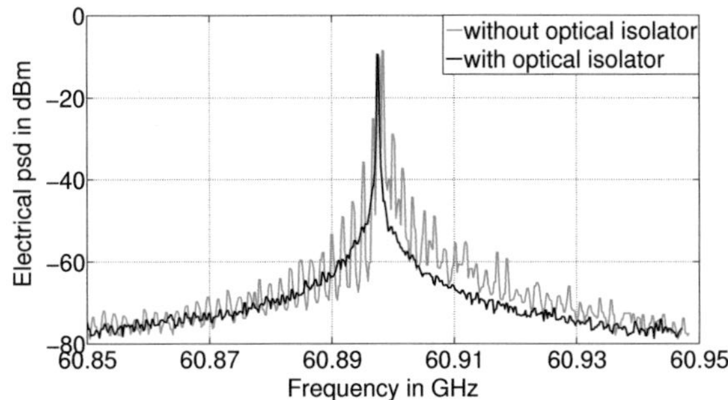

Figure 2.8: Electrical beat note with and without isolator (L872), RBW = 3 MHz.

of 0.6 dB. Figure 2.8 shows the effect of removing the isolator on the generated mm-wave signal. Reflections can be minimized using fibers treated for minimum reflections. In order to minimize optical coupling losses, lense systems or tapered micro-lense fibers can be used.

The transmission not only depends on the amount of light reflected backwards, but also on the effective coupling of light into the fiber. The crucial factor here is the so-called *mode field radius* w_0; its doubled value is often also referred to as *spotsize*. The intensity profile in an optical fiber is rotation-symmetric and can be approximated quite accurately by a Gaussian curve along the radial coordinate[14]. The mode field radius is defined as the margin where the power density decreases to a value of $1/e^2 = 13.5\%$ of the maximum. When the optical wave reaches the edge of the waveguide and continues to propagate into free space, the mode field radius increases in z-direction

$$w^2(z) = w_0^2 \left[1 + \left(\frac{z\lambda}{2n\pi w_0^2} \right) \right] \tag{2.29}$$

and is described by an angular distribution in the far-field

---

[14]The actual profile is given by v-order Bessel functions J_v in the core region and exponentially decaying Hankel functions in the cladding [Fis02].

$$w_0 = \frac{\lambda}{\pi n \tan \theta} \tag{2.30}$$

for Gaussian profiles and $w_0 \ll 1$ as well as $z \gg 100$ µm. For an efficient power coupling from chip to fiber, maximum overlap between the mode profiles should be achieved. For SMF, the magnitude of the overlap can be approximated using the ratio κ_0 as a function of the respective mode field radii w_0 of the fiber, and w_1 of the laser[15]:

$$\kappa_0 = \frac{4}{\left(\frac{w_1}{w_0} + \frac{w_0}{w_1}\right)^2}, \tag{2.31}$$

and the mode field mismatch $L(\kappa_0) = -10\log(\kappa_0)$ in dB ([Fis02], [SN79]).
The concept of mode field mismatch allows an estimation of the coupling losses on the assumption that laser and fiber are mechanically adjusted in an optimum way. The notion of *optimum* usually refers to a condition at which maximum optical power can be coupled from the laser into the fiber. However, optimum coupling in our particular context aims at another criterion, namely the maximum achievable electrical power of the fundamental beat note at best frequency stability and minimum phase noise. Frequency stability and phase noise characteristics of the beat note cannot be quantified in terms of $L(\kappa_0)$ as they depend to a large extent on the retro-reflections from the tip of the fiber into the laser cavity. Two coupling systems have been implemented in the testbench: An anti-reflection[16] coated tapered micro-lense fiber ($w_1 = 1$ µm$\pm$ 0.25 µm at a working distance of 70 µm) consisting of a rod lens that was fusion-spliced to an SMF was employed, as well as a dual lense focaliser system consisting of two aspheric lenses, a spot-size $w_1 = 3$ µm$\pm$ 1 µm at 0.9 mm working distance. Nominal return losses were 60 dB. Figures 2.9(a) and 2.9(b) show the MLLD air-coupled to the micro-lense SMF and to the focaliser, respectively.

---

[15]It is assumed that the field distribution in vertical and horizontal directions do not differ by more than a ratio of 1:3, otherwise both directions have to be considered separately.

[16]Anti-reflection coating is a technique which was originally conceived for the treatment of planar surfaces, such as lenses with small curvatures. There is evidence that the same technique is not quite as effective on highly bent surfaces, e.g. micro-lense fibers. This circumstance might explain the difference in performance of the two coupling systems.

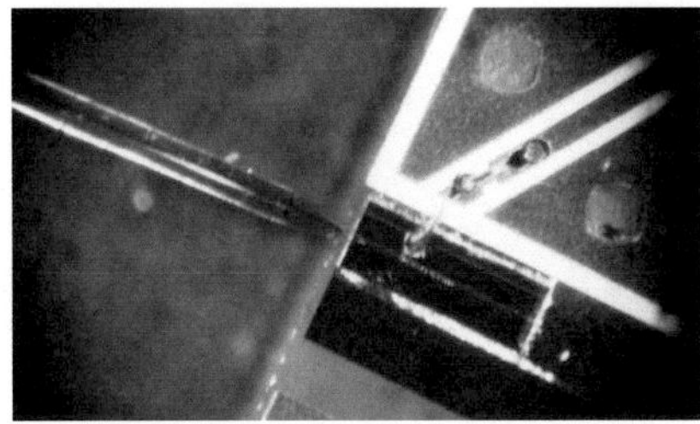

(a) MLLD with micro-lense SMF

(b) MLLD with focaliser

Figure 2.9: Coupling systems in the testbench

The micro-lense SMF or focaliser is placed on a movable sledge carried by a precision three-axis-positioner[17] in order to adjust the fiber such that optimum coupling is achieved. The mechanical stability is crucial for the long-term stability of the coupler setup; it is the main cause for the measurement uncertainty of as much as 1 dB in the optical domain. In particular, the set-up is sensitive to environmental influences such as heat and vibrations. In order to attenuate mechanical vibrations, the testbench is placed on an air-suspended table. In normal operation, i.e. above lasing threshold, the light emitted from the MLLD is approximately linearly polarized. This is due to the polarization dependence of the reflection coefficient at the emitting crystal facet, where the polarization vector is parallel to the longer side of the rectangular aperture. The light however loses its defined polarization when propagating through the lense-fiber systems and the pigtails of the optical isolator. At the output of the transmitter module, the MLLD thus delivers light of unknown polarization. In combination with external modulators, a polarization controller must be used to re-establish linear polarization and to couple a maximum amount of power into the modulator waveguide.

---

[17] from Melles-Griot

2.2.2 Properties of the mode-locked laser diode

In the following section, the laser characterization will be shown in static and in dynamic regimes, where the differentiation refers to configurations where the injection current is either a DC current I_B ("static"), or corresponds to a signal modulated around I_B at an intermediate frequency f_m ("dynamic"). In static regime, the average optical power, the shape of the optical spectrum, frequency stability, the beat linewidth at the fundamental mm-wave tone and the equivalent phase noise can be observed. The optimum bias current I_B is identified as the current yielding stable mode-locking and minimum phase noise. We shall look into relative intensity noise and introduce the chirp parameter. In dynamic regime, our foremost interest lies in the slope efficiency, the modulation response, the 3 dB-bandwidth achievable at the optimum bias point, and matching issues of the laser package.

2.2.3 Laser operation in static regime

For operation in static regime, a DC current is injected into the MLLD via the bias-T and the probe. The laser spectra are measured by means of an optical spectrum analyzer (OSA). The average optical power is measured using a sphere-based power meter. Figures 2.10(a), 2.10(c) and 2.10(e) show three exemplary spectra of the MLLD characterized in the course of this work. The spectra exhibit about 40 - 50 modes and are generally flat with a variation of 3 dB over 10 nm in the spectrum. The center wavelengths vary, and the fundamental beat notes are observed at frequencies ν_F ranging from 58.6 to 62.2 GHz. The frequencies ν_F are adjusted during the fabrication process by cutting the semiconductor structure to the appropriate length that lies in the order of about 700 μm. The precision of this technique is in the range of hundreds of MHz to several GHz. The key characteristics of the three diodes are listed in table 2.1.

2.2.3.1 Characteristic PI-curve and generated mm-wave carrier signal

Above the lasing threshold current I_{th}, the average optical power delivered by the laser is proportional to the amount of current injected into the laser before

Table 2.1: Parameters of MLLD characterized in this work, *) micro-lense SMF

Identifier	$FSR = \nu_F$ (GHz)	Center λ (nm)	$p_{RF,c}$ (dBm)	$p_{opt,max}$ (dBm)	Opt. I_B (mA)	I_{th} (mA)	max. s_1* (W/A)	3 dB BW (GHz)
L34	58.63	1556	-27	9	260	15	0.0215	2.10
L611	62.24	1529	-29	10	170	10	0.0206	1.00
L872	60.89	1573	-26	11.5	270	20	0.0129	3.80

reaching saturation for high pump currents. The power-over-current (PI)-curve so characterizes the emission of a semiconductor laser. The fiber-coupled PI-curves for the MLLDs are recorded at a temperature of 25° Celsius, see figures 2.10(b), 2.10(d) and 2.10(f). The lasing threshold is reached at around 15 - 20 mA. The fiber-coupled maximum output power using the micro-lense SMF ranges about 3 - 3.5 mW. The amount of coupled output power can be increased to about 10 mW using the dual lense system. The evolution of the RF power recovered in the fundamental beat note is not linear, but there exist discrete values of I_B that allow for stable mode-locking at relatively low phase noise. The optimum bias points are listed in table 2.1. Chip L34 shows best stability and is therefore selected for most transmission experiments. Chips L611 and L872 exhibit sharp drops in RF power at certain bias currents where mode-locking seems to be disrupted. The question arises as to whether the RF power in the fundamental beat note can reliably be predicted knowing the measured optical power, the shape of the spectrum, in particular the number of modes and the characteristics of the photo-detector. This is indeed possible within the limits of measurement uncertainty; table 2.2 sums up an analysis of the expected carrier power generated by the MLLD at $\nu_F = 58.6$ GHz, based on the power p_{opt} measured on the OSA. Taking into consideration a PD responsivity $\rho = 0.59$ A/W, i_{ph} is calculated to about 0.70 mA, representing $p_{el,tot}$ of -16.10 dBm delivered to 50 Ω. Based on the measured laser spectrum, the net loss of the low pass filter effect of the PD is calculated to -10.2 dB. From this derivation, we expect to find a carrier power of about -27.8 dBm.

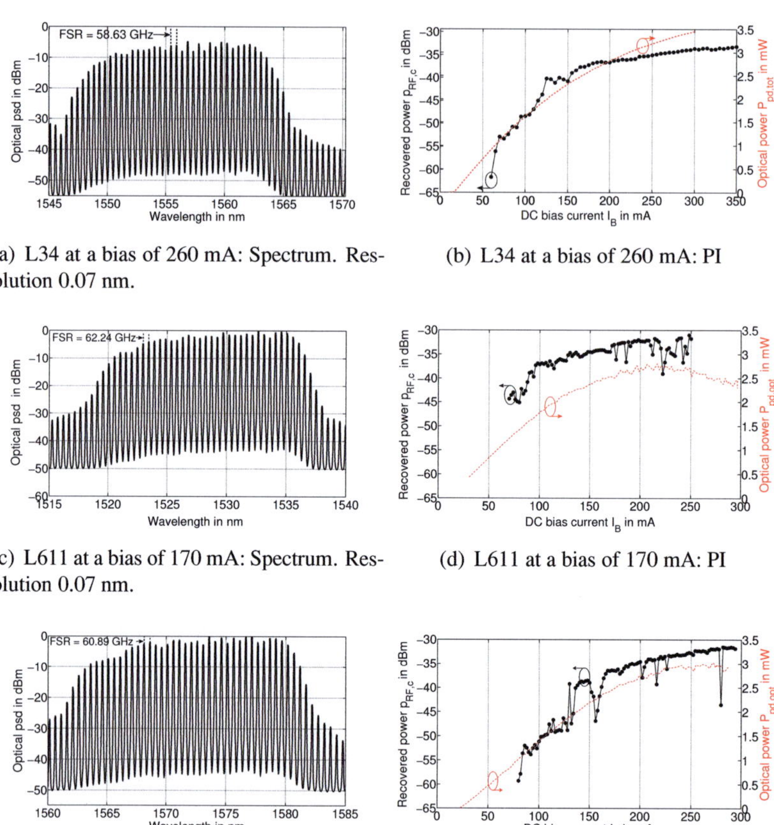

(a) L34 at a bias of 260 mA: Spectrum. Resolution 0.07 nm.

(b) L34 at a bias of 260 mA: PI

(c) L611 at a bias of 170 mA: Spectrum. Resolution 0.07 nm.

(d) L611 at a bias of 170 mA: PI

(e) L872 at a bias of 270 mA: Spectrum. Resolution 0.07 nm.

(f) L872 at a bias of 270 mA: PI

Figure 2.10: *Left column:* Measured optical laser spectrum. *Right column:* Static PI-curves (dashed curve, micro-lense SMF coupling) and recovered carrier power (solid curve with markers) with I_B. Selected results were previously published by the author in [BPCvD11a] and [BYP$^+$12].

Table 2.2: Calculated carrier power $p_{RF,c}$ from the optical spectrum,
L34 at optimum bias

Involved Powers	Power $\pm$ Uncertainty	
P_{opt} on PD (optical)	4 dBm	$\pm$ 1 dB
$p_{el,tot}$ in PD (electrical)	- 16.10 dBm	$\pm$ 2 dB
$p'_{RF,c,}$ in PD	- 26.30 dBm	$\pm$ 2 dB
RF loss due to mismatch of PD	- 1.5 dB	
Expected $p_{RF,c}$ at v_F = 58.6 GHz (in measurement: -27.3 dBm)	- 27.80 dBm	($\pm$ 2 dB)

Taking into consideration a significant measurement uncertainty (1 dB optical) originating from the mechanical coupling conditions, a carrier power of $p_{RF,c}$ = -27.3 dBm is achieved in the experiment under the condition of micro-lense SMF coupling which is reasonably close to the expected value.

2.2.3.2 Frequency stability and phase noise

Figure 2.11 shows the non-linear relationship between the fundamental beat frequency v_F and the bias current. A variation of about 20 MHz across the bias current range is observed. The error bars indicate the measurement uncertainty of the beat frequency, the manifestation of a slow frequency jitter over time which is visible to the naked eye (i.e. the rate of change is in the order of several Hz). The error bars are not drawn to scale but represent deviations of 200 kHz - 600 kHz around v_F. If, as in this case, the frequency of the carrier signal varies, it is common to speak of limited *long-term stability*.

In figure 2.12, we have plotted both a single-sweep spectral measurements of the lower sideband (resulting from modulating the MLLD-generated carrier with a sinusoidal signal at f_{IF} = 1.8 GHz) and an average curve of 50 subsequent sweeps to illustrate the effect of the frequency jitter on the spectrum when measured over time. The nominal frequency of the lower sideband is then 59.097 GHz for chip L872 ($v_F - f_{IF}$, v_F =60.897 GHz). Evidently, frequency jitter of that sort

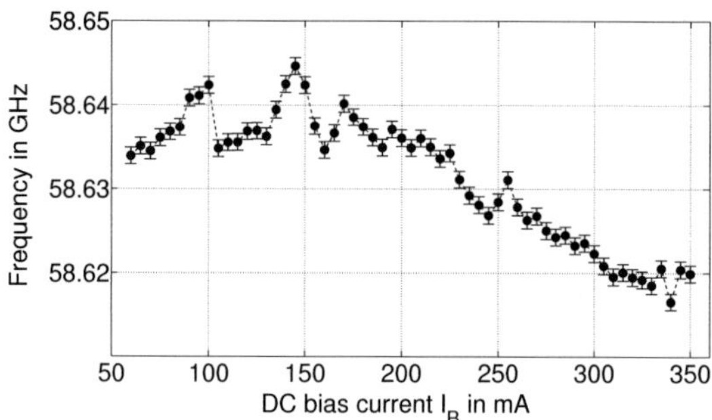

Figure 2.11: Frequency variation with I_B (L34)

is problematic as it demands a tracking of the carrier frequency upon reception in order to correctly demodulate the received signal. Such tracking is possible in a laboratory environment in which high-end radio receivers are employed for the purpose of testing. We will demonstrate transmission experiments in chapter 6 where appropriate receivers have been employed. Low-cost mobile devices however will not *per se* support this function. For standard-compliant RoF systems, where transmission channels are precisely defined and spectral masks have to be adhered to, the imprecision of the v_F selection is disadvantageous, as the intermediate frequency always needs to be adjusted accordingly. A synthesizer architecture incorporating the IF oscillator is therefore inevitable. The problem of frequency selection and frequeny jitter shall be picked up again in chapter 5, where stabilization architectures and methods of frequency synthesis will be discussed.

Figure 2.13 shows the fundamental beat note for chip L34 at a carrier frequency of about 58.63 GHz as measured with a resolution bandwidth (RBW) of 100 kHz at a span of 2.5 MHz. Due to the jitter of the carrier frequency v_F and the limited sweep time for a given RBW, the resolution could not be further improved. The measured linewidth $\Delta\widetilde{v}_{F,1}$ of about 100 kHz merely represents an upper limit to the actual linewidth of the quasi-Lorentzian spectrum, which is estimated to a few tens of kHz maximum [SGR$^+$]. As in the case of electrical oscillators, we speak of limited *short-term stability*.

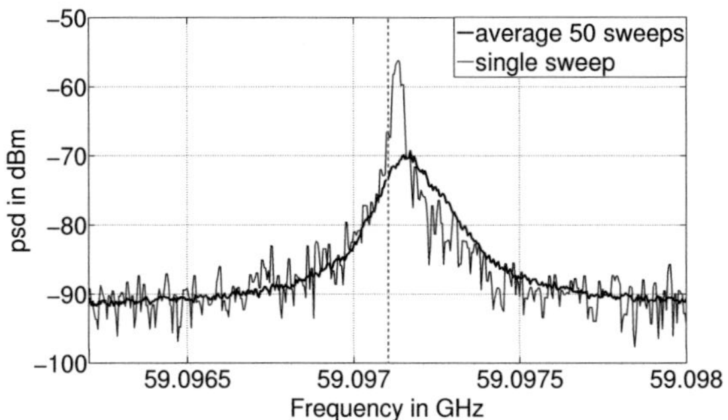

Figure 2.12: Measurements in the lower sideband under sinusoidal modulation at f_{IF} (L872), RBW = 10 kHz, I_{B} = 260 mA.

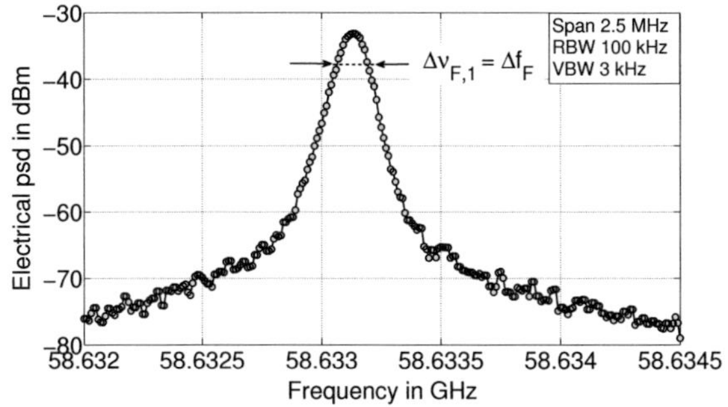

Figure 2.13: Measured beat note (L34), I_{B} = 260 mA.

In section 2.1.1.2 we saw how the oscillator spectrum is shaped by parametric noise, and in section 2.1.1.3 how the phase psd of the recovered mm-wave signal is the result of the different noise processes in the passively mode-locked laser. Except in the immediate proximity of the carrier, the $\zeta = 2$ noise process and the

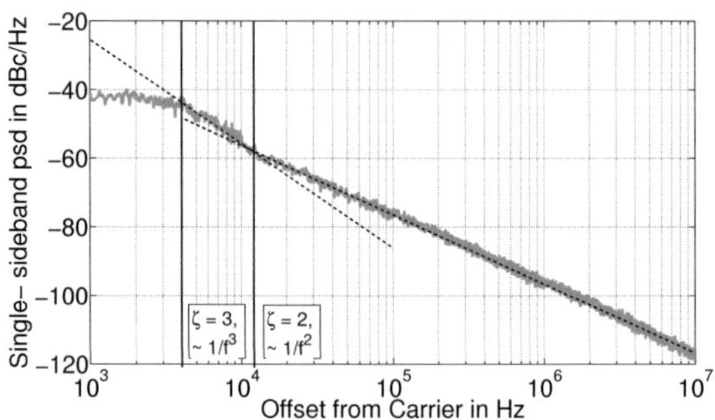

Figure 2.14: Typical phase noise performance of a free-running MLLD, from [vDEB$^+$08].

linewidth of the Lorentzian spectrum theoretically determine the psd of the RF phase noise ([KOT$^+$08] and references therein):

$$S_\varphi(f) = \frac{\Delta\nu_{F,1}}{\pi f^2}. \tag{2.32}$$

Figure 2.14 shows a typical SSB phase psd $L_\varphi(f)$ of the MLLD - a result previously published in [vDEB$^+$08]. Unfortunately, the measurement was not extended to sufficiently high offset frequencies for the $\zeta = 1$ noise process to show. From this curve, it is theoretically possible to determine $\Delta\nu_{F,1}$ via equation (2.32). Such an evaluation yields a linewidth of approx. 2 kHz for an MLLD stabilized in center frequency. This result represents, in turn, a lower limit to the linewidth and must be handled with care for three main reasons:

First, in figure 2.14 the $\zeta = 3$ noise process (flicker FM) can be identified for frequencies between 5 kHz and 10 kHz. The true linewidth must therefore be significantly broadened. Second, the instability of the carrier itself implies an uncertainty in the $L_\varphi(f)$ measurement that is hardly quantifiable; for frequencies lower than 5 kHz, the measurement is therefore not reliable. For the same reason, we do not observe the $\zeta = 4$ noise process in the measured phase psd. Third, the linear relationship between $S_\varphi(f)$ and $L_\varphi(f)$ which allows us to draw conclusions based on the measured psd only holds for small modulation by phase noise[18].

2.2.3.3 Relative intensity noise

Random emission of spontaneous and stimulated photons constitute a major contribution to the noise on the laser's optical power. Typically, the noise decreases with the average output intensity and is therefore referred to as *relative intensity noise* (RIN). The RIN is usually dominant over other forms of noise (e.g. shot noise) at the optical source [CI04]. As implied by its name, RIN is measured relative to the average optical power P_{opt} and specified in dBc/Hz. The effects of RIN can be observed at the electrical output of the photo-detector.

RIN measurements on chip L872 are shown in figures 2.15(a) and 2.15(b). In figure 2.15(a), RIN curves for bias currents from 50 mA to 250 mA, in steps of 50 mA, are shown. The measurement allows an evaluation of the relaxation frequency f_r of the laser at the maximum of the respective RIN curve. For high bias currents, it shifts towards 8 to 9 GHz. Typically, the RIN measurement also gives an estimate for the overall dynamic performance of the laser. However, the relationship between RIN and bandwidth measurements is not straightfoward for multi-mode lasers in general and for the MLLD in particular.

In figure 2.15(b) a measurement for 250 mA is shown in detail. A floor value of about -160 dBc/Hz was found. The peaks below 5 GHz are random parasitics which occur due to the electromagnetic environment; they could be disposed of by measuring the diode in a Faraday cage. Despite early predictions attributing a particularly low RIN level to quantum dot devices, the measured RIN performance of the quantum-dash MLLD is not significantly lower but comparable to those of quantum-well structures exhibiting low confinement factors [vD12].

2.2.3.4 Linewidth enhancement and chirp

In any semiconductor laser, the gain and the resonance wavelength are linked to the imaginary part n_I and the real part n_R of the refractive index.

---

[18] An integrated phase error $\ll 1\ \mathrm{rad}^2$ is usually required [SHR73].

Both quantities, or rather, their respective evolution with the concentration of charge carriers N in a pumped laser, are related to each other via the so-called *linewidth enhancement factor* or phase-amplitude coupling parameter[19] α_H [Hen82]:

$$\alpha_H = \frac{\frac{dn_R}{dN}}{\frac{dn_I}{dN}} \left(\frac{4\pi}{\lambda} \right). \tag{2.33}$$

The influence of the linewidth enhancement factor is reflected in the coupled differential equations known as the rate equations describing the laser's dynamics.

Generally speaking, the linewidth enhancement factor is a measure of the change in resonance wavelength when the laser pump level is varied, and in particular also when the laser diode is modulated. An ideal intensity modulator modulates only intensity, but any real device brings about a change in refractive index of the material under modulation so that some level of undesired phase modulation usually has to be accepted. The phenomenon is known as laser "chirping", by analogy to an audible linear modulation in audio frequency. The amount of parasitic modulation is then also characterized by α_H, in this context also referred to as *chirp parameter* [Zap04] which characterizes an intensity modulator. Ideally, an intensity modulator should have the lowest possible chirp value.

The passively mode-locked quantum dash laser emits chirped pulses not only when modulated but also under DC bias conditions. Experiments have revealed a so-called "blue chirp" for the passively mode-locked laser, referring to a situation where the instantaneous frequency of the pulse envelope increases with time, and the trailing edge of the pulse experiences an upshift in frequency [SS86]. Such a pulse is referred to as *positively* chirped (up-chirp), and it is an effect opposite to the chirping that occurs when a pulse broadens upon propagation on an optical fiber with (typically) positive group velocity dispersion, a subject studied in detail in chapter 4. The chirp phenomenon for MLLD has been analyzed in [STK96], where the blue chirping was attributed to the inclusion of a saturable absorber. In the case of the quantum-dash mode-locked laser, no such absorber section is present. *De facto*, we can observe a blue chirped pulse at the output of the MLLD that might be brought upon by negative GVD in the laser. However, the exact

---

[19] or "Henry's factor", a tribute to the author of [Hen82]

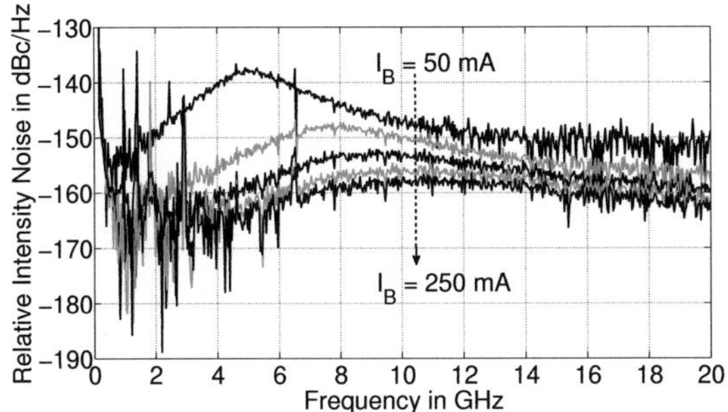

(a) RIN evolution with bias current

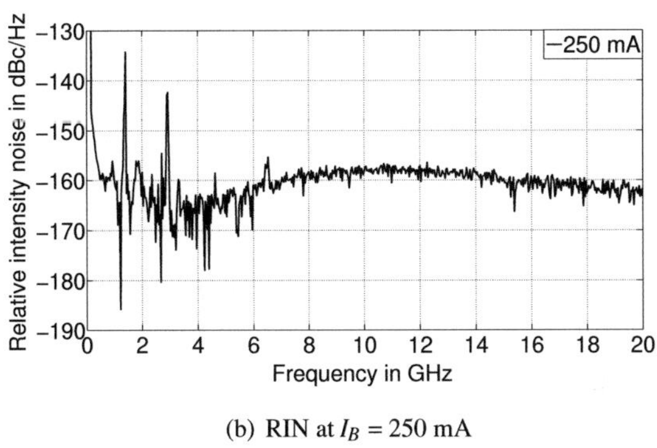

(b) RIN at $I_B = 250$ mA

Figure 2.15: RIN performance (L872): Courtesy of F. van Dijk, Alcatel Thales III-V Lab.

dynamics of the MLLD are still a topic of intensive study, and we can merely describe its behavior at system level.

As in the case of the single-mode laser, the chirp phenomenon relates the phases φ_m of the optical modes in the MLLD to α_H. In the case of the MLLD, $\varphi_m = f(\alpha_H)$ is a complex relationship due to the interaction of several non-linear processes between neighboring modes, and due to the gain curve associated with quantum dots of different sizes [GM07].

The chirp induces a deterministic phase shift between adjacent modes so that $\Delta\varphi_{m,m-1} \neq 0$ but constant, and the phase shifts $\Delta\varphi_{m,m-1}$ are altered through dispersion effects when the signal propagates in dispersive fibers. The RF power in the fundamental beat note - recovered after transmission across a given length of fiber - is directly affected. Through this phenomenon, the chirp affects the achievable propagation distances in as much as that it shifts the absolute distance where maximum power can be recovered. The phenomenon will be treated in detail in chapter 4.

Gosset *et al.* have demonstrated a method to characterize the amplitude and phase dynamics of a quantum-dash MLLD [GMM$^+$06]. The method is based on a Fourier analysis of the auto-correlation of several groups of three neighboring modes within the optical spectrum. The method could not be applied in measurement due to the absence of suitable filter devices. For a general idea, values of α_H as low as 2.5 have been measured for single-mode p-doped Q-dash lasers at ambient temperature [CLB$^+$10].

2.2.4 Laser operation in dynamic regime

In this section, the behavior of the MLLD under direct small-signal modulation at a frequency f_m shall be explored. The efficiency of the modulation in terms of the so-called *slope efficiency*, as well as its bandwidth and the impedance matching conditions, are investigated. These properties characterize the E/O conversion capabilities of the MLLD in a direct modulation scenario.

2.2.4.1 Slope efficiency

Electro-optical conversion of a time-variant analog electrical signal is achieved by modulating the laser's injection current around a bias point with this electrical signal. The corresponding PI curves have already been shown in figures 2.10(b), 2.10(d) and 2.10(f). In the absence of parasitic elements and substrate losses, the bias current is the current effectively pumping the active zone of the laser.

The upper limit to the performance of the electro-optical conversion process at a modulation frequency f_m is measured in terms of the *slope efficiency* of the characteristic curve at a bias point I_B in watts per ampere (W/A), ideally given by the laser quantum efficiency η_L, and the ratio of the energy of the incident photon, at a wavelength λ_0 at the speed of light c_0, to the electron charge q and the relaxation frequency of the laser f_r:

$$s_1(i_L = I_B) = s_{1,B} = \frac{\eta_L h c_0}{q \lambda_0} \cdot \frac{1}{1 + j\frac{\gamma f_m}{f_r^2} - \frac{f_m^2}{f_r^2}} \approx \left. \frac{d p_{opt}}{d i_L} \right|_{i_L = I_B}. \tag{2.34}$$

The slope corresponds to the approximate change in optical power proportional to a change in pump current. The following tangential approximation for optical power emitted from the laser will apply unless otherwise stated:

$$p_{opt} = s_{1,B}(i_L - I_{th}) = s_{1,B} i_L - s_{1,B} I_{th}. \tag{2.35}$$

The subtrahend $s_{1,B} I_{th}$ is a DC term; it will not be considered in the context of modulation efficiency as treated in chapter 3. The maximum achievable slope efficiency is wavelength-dependent even for a perfectly efficient stimulation (one photon per electron). At 1.55 µm, the maximum achievable slope efficiency that can be obtained at DC values is 0.8 W/A; typical slope efficiencies that have been measured are of the order of 0.1 W/A ([NMC$^+$93], [CUYBC95]). For the micro-lense-SMF coupled MLLD, slope efficiencies in the range of 0.02 W/A have been measured[20]; they are listed in table 2.1 in section 2.2.2. These relatively low values reveal the importance of efficient chip-to-fiber coupling.

For maximum slope efficiency, I_B is usually chosen in the linear region. However, we have seen that the amount of recovered RF power as well as the phase noise of the beat note also depends on the bias current. We thus encounter inevitable trade-offs between the three properties. Slope efficiency is a crucial factor in determining the modulation efficiency of the MLLD as observed in the laser's frequency response. Yet it is not the only one, as will be discussed in the next section.

---

[20] When the curve is measured on a packaged device, $d p_{opt}/d i_L$ refers to the single-ended fiber-coupled slope efficiency, which is the bare chip slope efficiency in series with the chip-to-fiber coupling efficiency. The double-ended slope efficiency would refer to an arrangement where light is collected at both facets of the device, which is not usually done.

2.2.4.2 Modulation response

The modulation response as quantified by the scattering parameter S_{21} of any laser is determined by its damping factor and its relaxation resonance, peaking at a relaxation frequency f_r. The two coupled sources of energy storage enabling this resonance are, on the one hand, the photons in the laser cavity, and on the other, the pool of charge carriers injected into the laser. Both photons and charge carriers have characteristic lifetimes τ inside the cavity, and their decay is due to the fundamental mechanisms of laser operation: stimulated emission and non-radiative recombination for the charge carriers, as well as emission out of, and absorption within, the cavity as well as spontaneous emission for the photons.

An increase in bias current implies an increase in injected carriers. In turn, a large number of carriers implies, in statisitcal terms, more opportunities for stimulated emission. The effective lifetime reduction resulting from this is responsible for shifting f_r to higher frequencies when the pumping is increased. A laser is therefore often operated at bias currents far above threshold and in the linear region of the PI curve [CI04], whereas we have seen that for the MLLD, the choice of bias current is determined primarily by the occurrence of stable mode-locking rather than by the consideration of the modulation response.

Figure 2.16(a) shows the modulation responses (S_{21}) of the MLLD for various bias currents from 50 mA to 250 mA, in steps of 50 mA. A relaxation resonance at 5 GHz for 50 mA can be observed; it is far less pronounced for higher bias currents. The modulation bandwidth of the MLLD is specified at the -3 dB point of the frequency response. The bandwidths measured at the optimum bias points for the different MLLDs range from 1 GHz to 3.8 GHz (see also table 2.1 in section 2.2.2). In the context of the RoF system, these bandwidths are considered IF bandwidths; up-conversion to the mm-wave range is achieved quasi-independently from the laser's modulation response.

In the vicinity of the mode-locking frequency, the modulation efficiency is increased due to the match with the optical cavity round-trip time. This phenomenon has been observed in measurement on similar structures and has been discussed in [vDEB$^+$08]. The effect can be seen in figure 2.16(b) for chip L872, where a peak appears in the response at the mode-locking frequency of 60.9 GHz. The response at low frequencies (about -40 dB) corresponds to the joint effect of

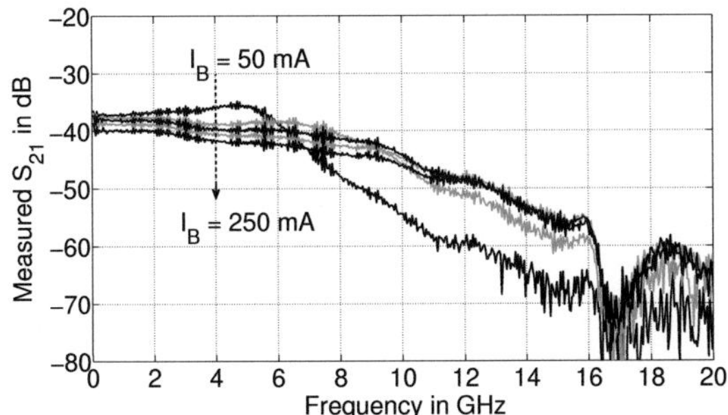

(a) Damping of the relaxation resonance for increasing pump currents

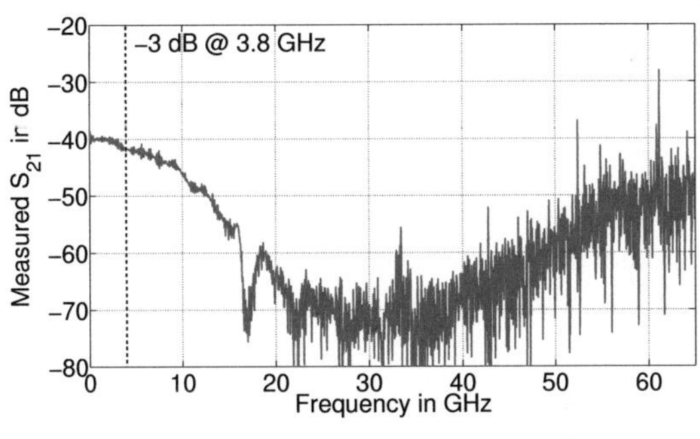

(b) Bandwidth measured at $I_B = 250$ mA

Figure 2.16: Bandwidth measurements (L872): Courtesy of F. van Dijk, Alcatel Thales III-V Lab.

slope efficiency, chip and package matching effects, chip-to-fiber coupling efficiency, and responsivity and matching of the photo-detector. While the slope efficiency, the chip-to-fiber coupling efficiency, the responsivity and even the matching of the detector[21] have little effect on the laser's frequency response, the

---

[21] within the specified frequency range of the detector

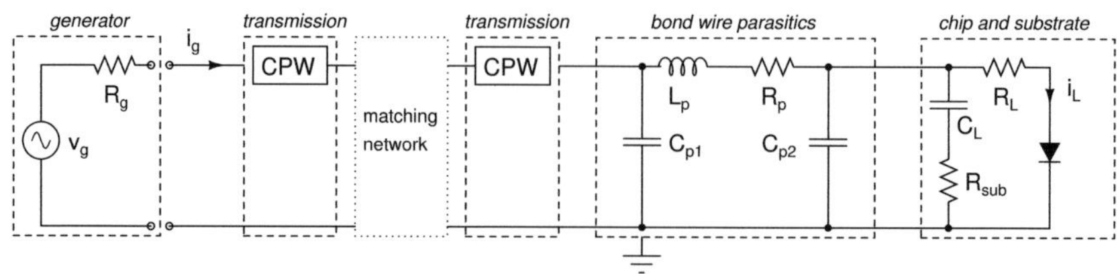

Figure 2.17: Circuit model of the laser diode including parasitics

chip and package parasitics do exert a considerable influence. In the following section, matching issues will be treated.

2.2.4.3 Impedance matching

Any effort to optimize the frequency response of the MLLD through improved matching requires detailed knowledge of not only the laser parameters but also of the actual implementation of the package and of the constraints of the assembly. Figure 2.17 shows the small-signal circuit model which reveals the different elements that come into play.

The laser itself can be described in an approximate manner by an ideal diode, its series resistance R_L, and its junction capacitance C_L, representing the influence of the carrier lifetimes. Substrate losses are reflected in R_{SUB}. The influence of the wire bond is represented by a π-network, consisting of the self-inductance L_p of the bond wire, the series resistance of the wire R_p, and the parallel capacitances C_{p1} on the die side and C_{p2} on the laminate side. The chip is accessed via a CPW line; its impedance is determined by its length and the substrate parameters (AlN). While in the design of the laser chip the CPW transmission line was laid out for a line impedance of 50 Ω[22], no further matching efforts have been undertaken during design. The real part of the laser impedance is approximately $R_L = 4.5 \ \Omega$. Other device or package parameters have not been determined. Resulting there-

---

[22] 50 Ω is the standard impedance that was available at the output of the vector signal generator. It is imaginable to design a vector modulator with an output impedance of less than 50 Ω. In this case, the mismatch between laser and signal generator could be reduced.

from, we find that the power reflected to the generator amounts to about 69.7% at low frequencies where the effect of the junction capacitance is yet negligible, and the bondwire parasitics have not been considered. A matching network can be inserted as shown in the figure. Several approaches have been explored in the past. They include:

- Resistive matching [SRD$^+$11]. The matching network consists of a simple series resistor. While the resistive match is independent of frequency, its fundamental problem is that a large portion of the electrical power will not be converted to optical power, but will be dissipated in the matching resistor instead.

- Reactive matching with stubs, or matching transformers. While stubs are typically used for narrowband matching [AMTO03], wider matching bandwidth can be obtained by employing multisection transformers, either consisting of lumped ([GK93]) or distributed elements, like step impedance transformer lines [GG90].

Techniques to improve the matching of the chip will be covered briefly in chapter 3, where direct modulation is discussed.

2.2.5 Limitations of MLLD sources

From the component-level characterization, we have identified four limitations of the MLLD that might eventually impede its use within the RoF link. These limitations are:

- **Frequency stability and phase noise.** In order to operate on a dedicated radio channel where a spectral mask has to be respected, the MLLD-based mm-wave generation must be able to place the desired band at a well-defined frequency, preferably with the lowest possible phase noise. However, the precision and the stability of the laser-based carrier generation is hindered by the inaccuracy of the frequency selection in the fabrication process of the MLLD, and by the frequency jitter of several hundreds of kHz which is present even at the optimum bias point. The single sideband phase noise of the MLLD was measured to about -55 dBc/Hz at an offset frequency of 10 kHz. At low offset frequencies, the phase noise cannot be measured due to the slow frequency jitter. From the measurement, we anticipate an important limitation on system performance, as phase noise and jitter generally influence the achievable quality of a communication system. Suitable stabilization architectures are necessary.

- **Power.** The primary interest that has governed much of the design of MLLD in the past has been the generation of optical combs (e.g. for source generation according to the ITU[23] grid) or very short pulses (e.g. in clocking or sampling applications). Device designers have therefore aimed at obtaining large spectra, as the spectral width is reciprocal to the pulse duration. For RoF applications, the opposite is true: in order to maximize the generated power in the electrical carrier, the spectrum should ideally be reduced to a minimum number of modes. More important still than carrier power is however the power the modulated MLLD can provide in a modulation sideband. We will see in chapter 3 that the consideration of the low-pass filter effect on the modulated spectrum will lead to the same conclusion.

- **Chip-to-fiber coupling.** When packaging laser diodes, designers have usually tried to achieve optimum coupling between chip and fiber in terms of maximum power coupling, i.e. in terms of the mode field mismatch $L(\kappa_0)$, see equation (2.31). For the mm-wave RoF link, maximum optical power is only of secondary interest; it is the power of the electrical carrier that should be maximized. Reflections on the fiber strongly influence the mode-locking; increased phase noise and reduced carrier power are the consequences. Minimizing the reflections from the surface of the fiber into the laser cavity should therefore be a priority during packaging, and reducing the sensitivity to external feedback should be considered a priority in laser design.

- **Impedance matching issues and bandwidth.** The modulation bandwidth and the matching of the MLLD are essential only in a direct modulation

---

[23] International Telecommunication Union

scenario. As was discussed above, the broadband matching of the MLLD is not a trivial task: even if the intrinsic modulation bandwidth of the diode theoretically allows for high-speed direct modulation, the implementation of the signal access on the diode chip is critical (bond wire parasitics etc).

Remedial measures that can be employed to reduce the detrimental effects on system performance include both the changes in the laser design, and the embedding of the device into a gradually more sophisticated transmitter architecture.

2.2.5.1 Device design

The problem of power can partly be addressed during device design: in terms of RF power delivered by the MLLD-PD pair, it was shown that maximum electrical carrier power can be obtained when reducing the number of modes to a minimum of three. This result was obtained when considering a situation in which the modes were perfectly coupled to each other and the optical power remained constant. It is generally possible to include a filter section on the laser chip that would largely reduce the number of lasing modes.

There is evidence that the coupling between the modes is strong only when a certain minimum number of modes is present, and will further strengthen with an increasing number of modes until reaching an optimum value where no further improvement is possible [vD12]. This indicates a probable trade-off between the number of modes - and thus, the generated electrical power in the fundamental beat note - and the phase noise characteristics of the device. The integration of a distributed Bragg reflector (DBR)[24] filter reducing the optical spectrum from 15 nm to about 8 nm in width has shown that the phase noise, quantified in terms of the beating linewidth of the fundamental note, can be maintained [vD12]. The reduction of modes is advantageous also when it comes to propagation effects like chromatic dispersion (see chapter 4). In combination with an external modulator and, eventually, an optical amplifier compensating for the insertion loss of

---

[24] A distributed Bragg reflector is a structure formed from various layers of alternating materials with varying refractive index resulting in periodic variation in the effective refractive index. The layer boundaries cause partial reflections of the optical wave. If the respective wavelength is close to four times the optical thickness of the layers, the many reflections interfere constructively. The DBR so acts as a high-quality reflector.

the modulator, it might still make sense to filter externally (again, see chapter 4). With respect to questions of stability (see below and chapter 5), it might be advantageous to introduce a saturable absorber in a second section.

2.2.5.2 Stabilization architectures

In conventional commmunication systems, efforts are usually undertaken to achieve frequency tunability and to improve the phase noise characteristics of the carrier. In all-electrical systems, the output frequency of a voltage controlled local oscillator is usually stabilized by a servo loop commonly referred to as phase-locked loop. Optical or opto-electronic phase-locked loops have been combined with mode-locked lasers. A phase-locked loop can be applied as long as the carrier signal is generated in such a way that it can be influenced by a controllable quantity (e.g. DC control voltage) so that small deviations from the nominal value can be compensated for by the loop. We have seen in the characterization that mode-locking occurs only for a certain set of pump currents. Therefore controlling the pump current offers only limited tuning capability. Unfortunately, the MLLD used in this work do not give access to other controllable quantities influencing the resonator, and there is no way to fine-tune the oscillation. Therefore, conventional phase-locked loops cannot be used. Other techniques to stabilize an MLLD include active mode-locking, optical injection, or external optical feedback. These techniques will be discussed in chapter 5, where we will propose an architecture which solves the problem of frequency instability and tuning for our particular application.

2.2.5.3 External modulators

Simplicity is the main advantage of direct modulation. This advantage, however, comes at the expense of matching problems, and a possible perturbation of the cavity by the injection of a modulated pump current. In order to circumvent the perturbation of the laser, and to avoid matching issues, the designer might consider the use of an external modulator. Where high-density packaging is not essential, the Mach-Zehnder interferometer configuration has served successfully. It is usually possible to match the modulator electrodes to 50 Ω (or a close value), so that impedance matching problems between the generator and the modulator can be avoided. The combination of the MLLD and the MZM will be an item of principal interest in the following chapter. A reader unfamiliar with this technique might want to consult appendix B, where the properties of the MZM are briefly presented. It is also possible to integrate an additional modulator section on the laser chip itself. Typically, electro-absorption modulators (EAM) are employed here.

2.3 Conclusions on chapter 2

In this chapter, the principle of operation of the mode-locked laser diode as well as a characterization of the diodes used in this work have been presented. The generation of mm-waves relies on the heterodyning of adjacent modes by means of a suitable photo-detector. While diode and photo-detector are, at least conceptually, simple devices enabling mm-wave generation up to hundreds of GHz, the challenge for device designers is to optimize the laser parameters with respect to the application. The device level characterization has revealed that for a given optical power delivered by the diode, a reduction of the number of modes is beneficial to the amount of carrier power generated from the MLLD-PD pair. For the modulated spectrum to be studied in the following chapter, we expect a similar result. A reduction of the number of modes is thought feasible during device design, although at the risk of further degrading the phase noise performance of the diode. Phase noise, or rather, the frequency jitter of the diode is another, and possibly more severe, limitation to the use of MLLD in mm-wave RoF systems. We maintain that if the problem of frequency jitter could be solved - preferably by a structure which would allow at the same time to improve the precision of the channel selection -, increased effort could be dedicated to limiting the spectrum on chip level.

No particular effort has yet been dedicated to efficiently match the laser diode to the electrical circuitry used for direct modulation. An estimated two-third of the available power is reflected back to the generator, and the intrinsic modulation bandwidth (at least 3.8 GHz) is not exploited. Electro-optical conversion of a modulation signal is thus expected to improve under proper matching conditions. MLLD matching will not be explicitly discussed in the following chapter, however, resources for further reading are indicated.

Concerning the packaging of the diodes, the experiments carried out in this chapter furthermore hint at the following: Chip-to-fiber coupling must be optimized for two parameters, i.e. optical power coupling and RF stability, which was found to largely depend on the level of optical retro-reflections from the fiber coupler into the laser. For the same reason, the package should include an optical isolator. In order to avoid any undesired influence on the mode-locking operation by the environment, the MLLD package should also include temperature regulation.

3 Analysis of the Mode-Locked Laser based Fiber Sub-System

While the preceding chapter was concerned with the component-level characterization of the MLLD as laser source for the analog link, this chapter will be dedicated to a system-level analysis of analog fiber links constructed on the basis of mode-locked laser-enabled signal generation and modulation. Ideally, the double conversion between the electrical and the optical domains is highly efficient, linear, independent from frequency and does not introduce additional noise or distortion. Needless to say, practical systems always to some extent fall short of these ideal characteristics, and the MLLD-based RoF link is no exception.

In contrast to a passive electrical link which can be defined in terms of its frequency response and loss only, an analog optical link rather resembles an active RF system. It therefore requires additional parameters in order to be fully described, namely output power, bandwidth, but also effective link gain, signal-to-noise ratio, noise figure, and dynamic range. These quantities are naturally inter-related; the reader is referred to the literature, e.g. [Ega03], for a thorough discussion of the relevant system concepts.

The quality of transmission a communication link can provide is typically measured in terms of the bit error rate (BER) of the digital signal (after the demodulation, the analog-to-digital conversion and the detection of a transmitted symbol), or in terms of the magnitude of the error vector of the analog signal (measured *before* the symbol is detected). Both BER and error vector magnitude (EVM) are related to the signal-to-noise ratio (SNR) of the analog signal. SNR is a crucial factor in determining BER or EVM, albeit not the only one, as we will discuss in chapter 6. In order to assess the transmission quality we can obtain from the MLLD-based RoF link, we will first need to determine the available SNR at the output of the link, i.e. the SNR which can be provided to the transmit antenna in the base station.

The SNR at the output of the link is determined by the link gain, but also by the amount of noise added by the components in the link, commonly quoted as the link noise figure. At the output of the link, it is, by definition, the ratio of the power of the electrical signal p_{out} to the sum of all detected noise powers, represented by σ_{d}^2, delivered at the photo-detector to a load R_0,

$$SNR_{\text{out}} = \frac{p_{\text{out}}}{\sigma_{\text{d}}^2}. \qquad (3.1)$$

Other than SNR, which is a relative metric, we must also be interested in the absolute output power the link can provide. For a given input power p_{in} the numerator of equation (3.1) is

$$p_{\text{out}} = p_{\text{in}} \cdot g, \qquad (3.2)$$

where g is the gain of the MLLD-based link. Typically, g is <1.

Link gain, noise and distortion have contributions from the different link components, including in particular the MLLD and the photo-detector, as well as external modulators, or amplifiers. In this chapter, two downlink architectures employing different modulation techniques are compared in terms of their system parameters. In a direct modulation link, the MLLD serves simultaneously as mm-wave signal generator and as E/O modulator. In an external modulation link, the MLLD only generates the mm-wave signal necessary for up-conversion. E/O conversion is achieved by means of an external modulator - in our case, an MZM. Fundamental trade-offs between the different modulation techniques exist and will be identified in the following.

3.1 System analysis of the MLLD-based link

In the analog fiber link, two basic operations can be distinguished: electro-optical conversion, where a time-variant modulation signal is converted into a time-variant intensity modulation of the laser light; and frequency conversion from IF range to mm-wave range, **if** E/O conversion cannot directly be achieved at mm-wave frequencies. This frequency conversion - which effectively corresponds to a mixing process - is realized through optical heterodyning, as discussed in chapters 1 and 2. For the MLLD-based link, these operations are illustrated in figure 3.1, corresponding to the downlink[1].

We have seen in chapter 1 that the downlink of the RoF system, where up-conversion to the mm-wave range is necessary, is best implemented as RF-over-fiber link where the transmit signal is directly available at the output of the photo-detector. In such an architecture, the MLLD provides the LO for up-conversion to the mm-wave range by mode-mode beating. E/O conversion can be achieved directly, i.e. at the MLLD, or externally, using an external modulator. When the MLLD is directly modulated, as shown in the upper part of figure 3.1, it performs both functions at the same time: mm-wave signal generation is obtained simply by biasing the laser at an optimum injection current I_B; electro-optical conversion is then achieved by modulating the injection current around I_B. In the external modulation link, the MLLD provides at its output a non-modulated signal that is subsequently modulated in its intensity by an electro-optical modulator. It is thus the voltage of the modulator which is varied around a bias point V_B in the transfer curve. At the output of either link, the signal is O/E converted and radiated across the air interface by means of a transmit (Tx) antenna.

3.1.1 Generation of the electrical DSB spectrum through mode-sideband mixing

When the optical field of the MLLD is modulated at an intermediate frequency f_{IF}, sidebands around each optical mode are created at the frequencies

$$v_m \pm f_{IF} = (v_0 + m \cdot v_F) \pm f_{IF} \tag{3.3}$$

where v_0 is the frequency of the first mode of the comb. In the upper part of figure 3.2, an exemplary optical spectrum of three modes is shown, where the lower and upper optical sidebands are labeled "L" and "U", respectively. This double sideband spectrum repeats itself for about 30 to 40 strong optical modes

---

[1] "Uplink" and "downlink" indicate the respective directions of signal transmission. They are not to be confused with the notions of "up-conversion" and "down-conversion" which refer to shifting the signal in the frequency domain through mixing. The downlink must operate at mm-wave frequencies, i.e. up-converted signals.

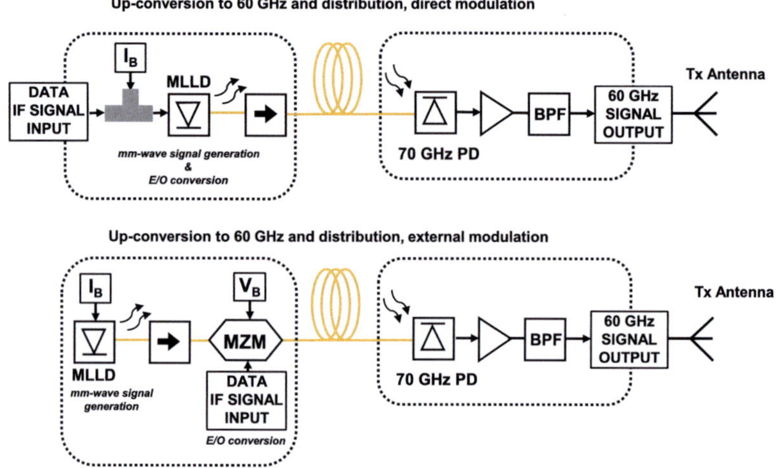

Figure 3.1: MLLD used in the mm-wave downlink. The information is sent "downwards" from the network to the user (CS to BS to MT).

in a real device. In chapter 2, we have shown how the mutual interference of all modes can be exploited for mm-wave carrier generation. In a similar way, a modulation brought onto the optical signal can be recovered in the electrical domain as a double sideband spectrum around the carrier. The electrical sidebands are created through simultaneous mode-sideband beating and the superposition of all corresponding beat signals in the electrical domain. Each optical mode will beat with the closer optical sideband of an adjacent mode to create the lower electrical sideband. Each optical mode will also beat with the farther optical sideband of an adjacent mode to give the upper electrical sideband. In principle, this process can be understood as a mixing operation between the different components. In figure 3.2, these different contributions are shown separately. The question might arise as to whether this form of mixing between a mode and a modulation sideband produces, at any moment, a phase reversal that will lead to a superposition of sidebands in phase opposition. A (partial) cancellation of (some of the many) beat signals might then lead to an electrical sideband in counter-phase to the actual one. As a result, a phase reversal of the detected symbols might occur which would not produce a high value of EVM but a high BER.

It can however be shown that the result of the mixing process is correct. We recall that in a mixer, the sideband at a frequency f_{RF} to be down-converted will keep its normal shape, or flip over to its inverse position, depending on the relative position of the local oscillator at a frequency f_{LO}. For $f_{LO} > f_{RF}$, the sideband is reversed, while for $f_{LO} < f_{RF}$, the normal frequency position is maintained. If we now consider the mode-sideband beating in the MLLD separately for each pair of spectral components involved, the respective mode acts as LO in the down-conversion of a given sideband. For the three-mode spectrum, this principle is illustrated in figure 3.2: the upper four drawings correspond to the beatings of a mode and a close-by sideband, resulting in the lower electrical sideband, whereas the lower four drawings correspond to the beatings of a mode and a farther sideband, resulting in the upper electrical sideband. It can be observed from the figure that the flipping-over to the inverse position of two out of four sidebands is required to assure the coherent superposition of the four sidebands at a time. Thus, a systematic phase error does not occur. This consideration applies as long as all the modes and the sidebands of the optical spectrum are in phase. In chapter 4, we will see that the situation presents itself differently when the different spectral components experience a phase shift upon propagation across the dispersive fiber.

3.1.2 Principles of intensity modulation

For the cost-effective implementation of the MLLD-based link, aspects like size and cost usually speak in favor of direct modulation. On the other hand, external modulation implies its own less obvious advantages, such as the fact that the locking of the modes will not be disturbed by injecting a modulation signal into the laser cavity. A comprehensive comparison of both modulation techniques must be made in terms of their respective gain values, noise figures, and distortion characteristics. We will derive analytical expressions for these system parameters of the MLLD-based RoF link and compare our findings to the measured values, a summary of which is provided in table 3.2 at the end of section 3.2.3. We will

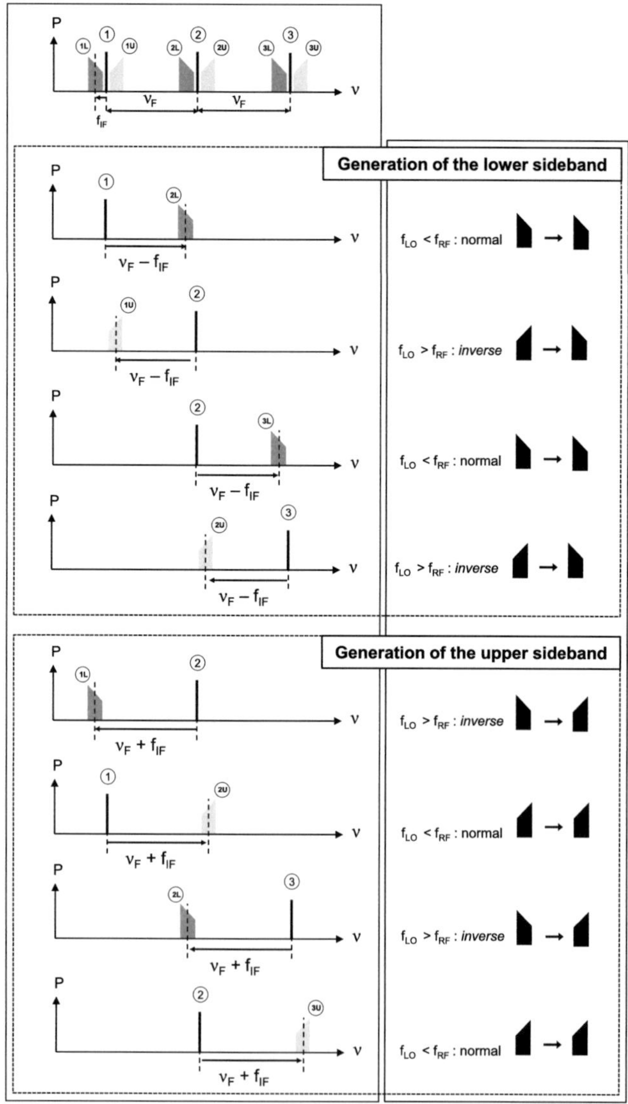

Figure 3.2: Sideband mixing, multimode spectrum

also identify techniques to improve system performance by optimizing the link parameters.

Physically, both direct and external laser modulation imply a modulation of the intensity of the light out of the laser or modulator device. Direct IM of the laser is achieved through a small-signal modulation around a pump current I_B. From the PI curve in figure 3.3(a) we observe that an analog variation of the pump current i around a DC value I_B results in an analog variation of the optical power p_{opt} around an average optical power value $\overline{P}$. From the figure, we can already observe a particular quality of the PI curve which we will study in detail in section 3.2.1.1: the steeper the PI curve, the greater the power variation for a given variation of current, and the more efficient[2] the E/O conversion by direct modulation.

External modulation relies on a different physical effect. In a lithium niobate ($LiNbO_3$) crystal MZM, the amount of optical power transmitted through the modulator varies periodically with the bias voltage V_B on its electrodes. MZM operation relies on two elements: first, the Pockels effect, namely, the effect of changing the refractive index of the crystal - and therefore, the phase retardation of a lightwave propagating through the crystal - by applying an electrical field, and second, the interferometric structure of the waveguides transforming a phase change into a change in output intensity. More details on the Mach-Zehnder modulator can be found in appendix B. For linear operation, V_B must be chosen at the quadrature point in the linear region of the MZM transfer curve as shown in figure 3.3(b). Again we observe that the efficiency of the E/O conversion depends on the steepness of the transfer curve at the bias point. The relevant parameter in this case is the so-called *half-wave voltage* V_π which is the voltage required to induce a phase shift of $180°$ in one of the two branches of the interferometer. For maximum efficiency, V_π is as small as possible.

For the following comparison of direct and external modulation, low-index[3] single-tone modulation with a sine wave is considered. The use of a sine wave is advantageous as it is manageable both analytically and experimentally, and it allows

---

[2]Here the notion of "efficiency" refers to the conversion between electrons and photons in the laser process rather to a power efficiency. In order to properly describe the latter, we would need to consider the laser input impedance.

[3]The modulation depths for analog applications are typically small and therefore in sharp contrast to the modulation depths in digital application, which typically approach 100% [CI04].

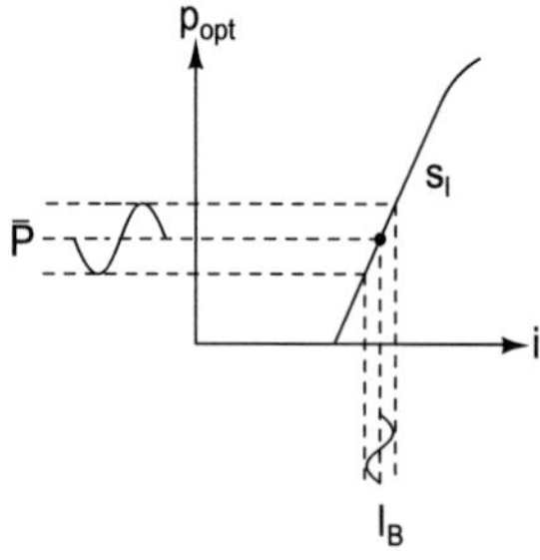
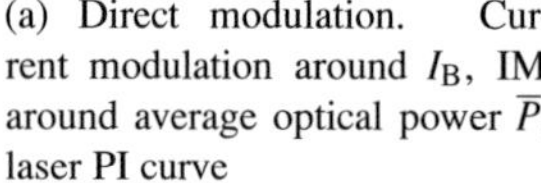

(a) Direct modulation. Current modulation around I_B, IM around average optical power $\overline{P}$, laser PI curve

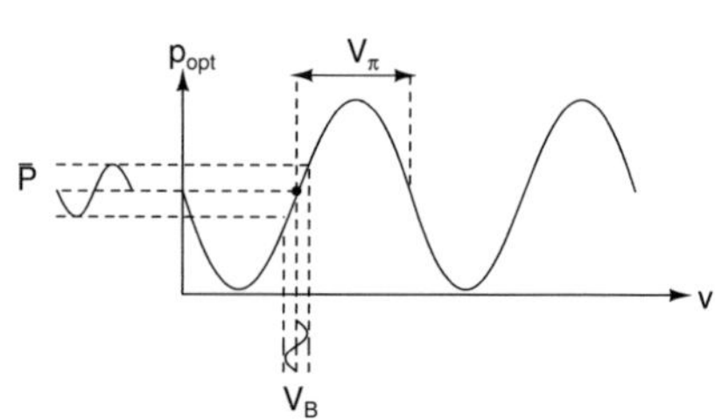

(b) External modulation. Voltage modulation around V_B, IM around average optical power $\overline{P}$, modulator transfer curve

Figure 3.3: Principles of intensity modulation

for a generalization of the results since any other signal can be decomposed into a sum of sine waves. In the following section, the measurement setup for direct and external link modulation will be presented. Some of the results have been previously published by the author in [BPC$^+$11a], but a more thorough discussion is presented here. We will first present the experimental implementation of the two links before introducing the relevant theoretical concepts. We can then directly compare analytical and experimental findings.

3.1.3 Experimental implementation of the MLLD-based downlink

In figure 3.4, the two RoF transmitters (at the CS) denoted with "Tx" are shown. In the upper part, the direct modulation link is represented. The MLLD is placed on the testbench previously described in section 2.2.1, and the light is collected using a micro-lense SMF[4]. The modulation source is a single-tone power signal generator, or an ensemble of two isolated signal generators for the intermodula-

---

[4] At the time of the measurements, the focaliser coupler was not yet available.

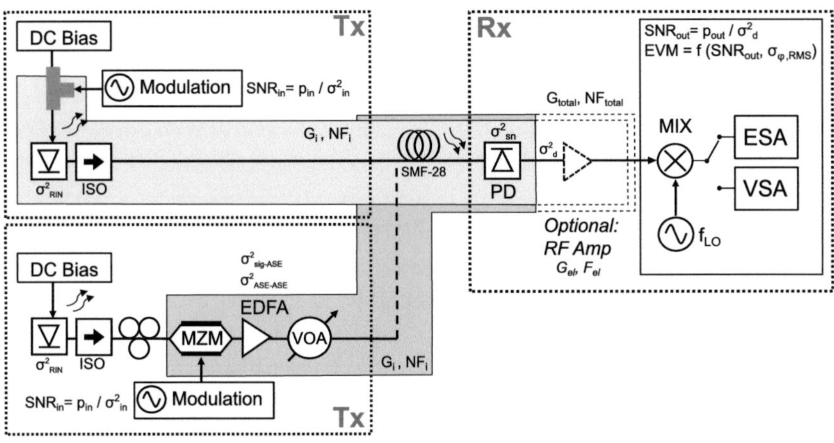

Figure 3.4: Downlink test setup for direct and external modulation

tion tests. The signal is then sent across an SMF[5] link whose length was chosen with regard to the dispersion optimum (a discussion of link dispersion will follow in chapter 4). In the lower part of the figure, the external modulation link is shown. This transmitter architecture includes several additional components, notably a single-electrode external MZM.

Maximum input power is coupled into the MZM[6] when the incoming light is polarized in parallel to the extraordinary optical axis (z-axis) of the $LiNbO_3$ crystal. At the output of the laser, the MLLD light is approximately linearly polarized. Upon propagation through a (polarization non-maintaining) fiber, this property is lost. At the input of the MZM, the polarization is therefore *a priori* unknown. A polarization controller is necessary to re-establish linear polarization and thus assure optimum power coupling of the optical signal into the modulator.

An erbium-doped fiber amplifier (EDFA)[7] is included in order to compensate for

---

[5] Standard SMF-28

[6] We used a dual electrode T-DE1.5-10-S-FN Mach-Zehnder modulator manufactured by Sumitomo Osaka Co., Ltd. in a single-electrode configuration.

[7] The device was manufactured by Manlight and exhibits a gain of ≈ 21 dB with a pump source at 980 nm.

the modulator's insertion loss. Due to the large gain of the EDFA and the optical power limitation of the PD, a variable optical attenuator (VOA) is employed in order to protect the diode from excessive optical powers. The VOA is set such that the average optical power on the PD remains the same as with direct modulation. Thus, the PD is operated under the same detector noise conditions for both configurations.

The RoF receiver (at the BS, denoted with "Rx") consists of a p-i-n photodiode[8] with a cut-off frequency of 70 GHz and an optional electrical amplifier[9]. The photodiode is a packaged device matched to 50 Ω at the output V-band coaxial connector. In the electrical domain, the signal can be observed on an electrical spectrum analyzer (ESA) that allows us to measure SNR, or in the case of data transmission (see chapter 6), on a vector signal analyzer (VSA) which can perform signal demodulation and EVM calculation. Both ESA and VSA have an input bandwidth limitation (40 GHz and 6 GHz, respectively), which is why the electrical mm-wave signal is down-converted to IF range using a broadband mm-wave mixer[10].

In figure 3.4, we can observe the relevant system parameters, such as the link gains g, the system noise figures NF, the different noise powers σ^2 arising at different points of the system, and input and output SNR. In the following discussion, the symbols used refer to this figure.

3.2 Comparison of modulation techniques

In order to assess direct and external modulation links, we will first look at how much gain or loss we can expect from the MLLD-based RoF link employing the respective modulation technique. The parameter which we will base this analysis on is the *intrinsic link gain*, i.e. the transducer power gain of the analog optical link without amplification. The concept of intrinsic link gain is based on

---

[8]We used a 70 GHz photo-detector manufactured by U2T with a responsivity of $\rho = 0.59$ A/W.

[9]We used the Terabeam HXI HLNAV 361.

[10]We used the broadband down-converter M60-5 from Spacek Labs.

the notion of *incremental* efficiencies[11] as they were first introduced by Cox, III [CI04]. Intrinsic link gain is the crucial factor for determining output power, but it also exerts a strong influence on link noise figure and SNR. Therefore, we are normally interested in maximizing this parameter. In the following, we will first determine the different contributors influencing intrinsic gain and later present methods of improvement by optimizing the contributing terms.

3.2.1 Intrinsic link gain and PD low-pass filter effect

In an all-electrical link, cascading components is not trivial, as each additional component represents a load to the preceding circuit that must be accounted for. Therefore, components are usually designed for a standard impedance that would allow the simple construction of a cascade. In an optical link, however, the different stages can be considered isolated in terms of load impedance. As a consequence, the link response is simply the product of the response of the different components, e.g. the modulation and photo-detection circuits. The intrinsic link gain g_i from the source with an available power of $p_{g,a}$ and into a load R_0 can be described as [CI04]:

$$g_i = \frac{p_{\text{load}}}{p_{g,a}} = \frac{p_{m,\text{opt}}^2}{p_{g,a}} \cdot T_L^2 \cdot \frac{p_{\text{load}}}{p_{d,\text{opt}}^2} \cdot T_b, \qquad (3.4)$$

and

$$G_i = 10 \log(g_i) \qquad (3.5)$$

is its value in decibels. The parts of the composite term g_i will be considered separately:

The first ratio $p_{m,\text{opt}}^2/p_{g,a}$ relates the available power of the generator to the amount of modulated power in the optical signal $p_{m,\text{opt}}$. It is referred to as the incremental modulation efficiency, a figure of merit that characterizes the optical source. The unit of incremental modulation efficiency is watts. The choice of modulation technique is crucial when determining its value.

---

[11] In contrast to the common notion of an "efficiency" which is dimensionless, incremental efficiencies can carry a unit that reduce to "1" when put in a chain.

The optical transmission loss T_L is the loss the signal suffers upon propagation; it is generally low for short-haul links. We assume that the link length has been optimized for dispersion effects (see chapter 4). (T_L^2) in decibels is then about -0.03 dB for SMF and at center wavelengths around 1550 nm. (T_L^2) might also contain losses of optical connectors or couplers.

The second ratio represents the incremental efficiency of the detection, i.e. the figure of merit of the PD given by its responsivity ρ and the load impedance R_0:

$$\frac{p_{\text{load}}}{p_{\text{d,opt}}^2} = \rho^2 R_0. \tag{3.6}$$

The unit of the incremental detection efficiency is, by definition, 1/watts. This term remains the same regardless of the modulation technique.

The last variable, T_b, reflects the particularity of the MLLD-based link, namely the fact that only a certain fraction of the input modulation power p_{in} will actually be recovered in the detected electrical sideband. The derivation of T_b requires some preliminary considerations. When all optical modes are simultaneously modulated, p_{in} is shared by all optical modes. Through photo-detection, it transfers to all electrical beat notes. In chapter 2, it was demonstrated that the band-limited photo-detector effectively filters the beat spectrum so as to recover only the fundamental beat note at a frequency ν_F. The electrical sidebands originating from the beating of their optical equivalences are affected by the same filter effect. The modulation power transferred to the sidebands around the higher harmonic beat notes represents a net loss, independent from the efficiency of the E/O conversion. This loss is reflected in T_b - which then represents the fraction of p_{in} recovered in one of two electrical sidebands.

We can estimate the magnitude of T_b by making the following consideration. We assume symmetrical sidebands on both sides of the respective optical mode. For intensity modulation, the current i_L flowing through the active zone is

$$i_L(t) = I_B \cdot (1 + i_m(t)) = I_B \cdot (1 + m_i \cos(2\pi f_m t)), \tag{3.7}$$

with the DC bias current I_B and the single-tone current modulation signal $i_m(t)$ at a frequency f_m where m_i is the modulation index. With equations (2.19), (2.20),

(2.21), (2.35), and (3.7), the detected photocurrent is

$$
\begin{aligned}
i_{\mathrm{ph}}(t) &= \rho \cdot s_{1,\mathrm{B}} \cdot I_{\mathrm{B}} \cdot (1 + i_{\mathrm{m}}(t)) \\
&= \frac{1}{4} \cdot \rho \cdot c \cdot \varepsilon_0 \cdot A \left(\sum_{m=1}^{M} |A_{\mathrm{m}}|^2 \right) \\
&\quad + \frac{1}{2} \cdot \rho \cdot c \cdot \varepsilon_0 \cdot A \left(\sum_{m=2}^{M} \sum_{n=1}^{m-1} |A_{\mathrm{m}}| |A_{\mathrm{n}}| \cos\left(2\pi(m-n)\nu_{\mathrm{F}}t + \Delta\varphi_{\mathrm{m,n}}\right) \right) \\
&\quad + \frac{1}{4} \cdot \rho \cdot c \cdot \varepsilon_0 \cdot A \cdot m_{\mathrm{i}} \left(\sum_{m=1}^{M} |A_{\mathrm{m}}|^2 \cos\left(2\pi f_{\mathrm{m}}t\right) \right) \\
&\quad + \frac{1}{2} \cdot \rho \cdot c \cdot \varepsilon_0 \cdot A \cdot m_{\mathrm{i}} \left(\sum_{m=2}^{M} \sum_{n=1}^{m-1} |A_{\mathrm{m}}| |A_{\mathrm{n}}| \cos\left(2\pi((m-n)\nu_{\mathrm{F}} - f_{\mathrm{m}})t + \Delta\varphi_{\mathrm{m,n}}\right) \right) \\
&\quad + \frac{1}{2} \cdot \rho \cdot c \cdot \varepsilon_0 \cdot A \cdot m_{\mathrm{i}} \left(\sum_{m=2}^{M} \sum_{n=1}^{m-1} |A_{\mathrm{m}}| |A_{\mathrm{n}}| \cos\left(2\pi((m-n)\nu_{\mathrm{F}} + f_{\mathrm{m}})t + \Delta\varphi_{\mathrm{m,n}}\right) \right).
\end{aligned}
\tag{3.8}
$$

The first summand is the DC contribution to the photocurrent while the second corresponds to the beat notes of the MLLD, i.e. the generated mm-wave carrier at frequencies ν_{F} and its harmonics at frequencies $m \cdot \nu_{\mathrm{F}}$. The third summand corresponds to the modulation signal at f_{m} which remains present in the detected RF spectrum. Finally, the fourth and the fifth summands correspond to the modulation sidebands on the generated mm-wave carrier at frequencies $\nu_{\mathrm{F}} \pm f_{\mathrm{m}}$ and to its harmonics at $m \cdot \nu_{\mathrm{F}} \pm f_{\mathrm{m}}$. The modulation power[12] p_{in} is therefore distributed among the third, fourth and fifth summands.

The photocurrent from equation (3.8) is effectively filtered by the band-limited detection, and the resulting double sideband signal will be band-pass filtered again when one sideband is chosen for transmission. If we relate the power in one of two sidebands around ν_{F} (fourth **or** fifth summand, and $n = m + 1$) to the total modulation power present in the sidebands around ν_{F} and its harmonics

---

[12]Here, the modulation power delivered to the laser active zone is considered. In the case of ideal matching, it equals the available power of the modulation generator.

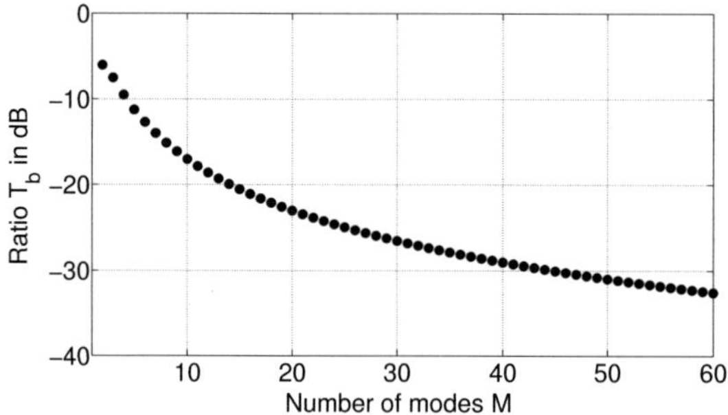

Figure 3.5: Ratio T_b, single sideband power to overall modulation power, induced by the filter effect on a selected sideband

(third **and** fourth **and** fifth summands), T_b is:

$$T_\mathrm{b} = \frac{\left(\sum_{m=1}^{M-1} |A_\mathrm{m}|\,|A_\mathrm{m+1}|\right)^2}{2\left(\sum_{m=2}^{M}\sum_{n=1}^{m-1} |A_\mathrm{m}|\,|A_\mathrm{n}|\right)^2 + \frac{1}{2}\left(\sum_{m=1}^{M} |A_\mathrm{m}|^2\right)^2}. \tag{3.9}$$

Note that in a DSB spectrum where only one sideband is selected for transmission, there is always a minimum loss of 3 dB (factor of 0.5). However, T_b also takes into account the power transferred to the original modulation frequency and $T_{\mathrm{b,max}} = 0.4$ for $M = 2$. Fig. 3.5 shows the loss that a single sideband will suffer from the PD filter effect. Maximum power is obviously recovered for M = 2. The loss indicated in the figure represents the net loss inherent in the optical generation of the sidebands by the MLLD, regardless of the modulation technique. T_b cannot be improved through improved slope efficiency (i.e. an increase in p_opt for a given I_B) nor through matching networks, but only through reducing the number of lasing modes in the MLLD for a given amount of optical power.

This scenario is true for a perfectly flat spectrum. In the case of the MLLD used herein, about $\geq 95\%$ of the optical power is concentrated in the inner 30 modes. From figure 3.5, a power loss of approximately 26.5 dB must be taken into account for calculating the sideband power P_{SSB} in a single sideband selected from the DSB spectrum after the PD.

In links based on the modulation of a single laser mode, positive intrinsic link gain can be achieved in both direct and external modulation systems ([CIRRH98], [ABB$^+$05]). In the MLLD-based link, however, the intrinsic link gain must always be <1 (negative in dB), unless T_b can be compensated for. This is only feasible in an external modulation link with amplification and filtering: we could e.g. use a combination of an optical amplifier, an optical filter and an external modulator in order to boost signal power at the output of the MLLD, then filter a small number of modes, and finally modulate the filtered spectrum.

3.2.1.1 Intrinsic link gain for direct modulation

Intensity modulation of a directly modulated laser is always accompanied by a frequency modulation of the laser signal. Olesen and Jacobsen have derived a general description of modulated laser fields and power spectra for a directly modulated single-mode laser [OJ82]. Their findings can be extended to the MLLD: with equation (2.4) and neglecting the dependence on z, the modulated laser field is

$$e_{m,opt}(t) = m(t) \cdot e_{RF}(t) \cdot \exp(j2\pi v_0 t), \qquad (3.10)$$

where $m(t)$ is the field modulating function. For sinusoidal modulation of the form

$$i_m(t) = I_B + I_m \sin(2\pi f_m t), \qquad (3.11)$$

the field modulation function is

$$m(t) = \sqrt{(1 + m_i \sin(2\pi f_m t))} \cdot \exp(j\beta \cos(2\pi f_m t) + \phi_f(I_B, f_m)), \qquad (3.12)$$

where $m_i = (I_m)/(I_B - I_{th})$ is the IM index, depending on the laser bias current, the amplitude of the modulation signal, and the laser threshold current; and

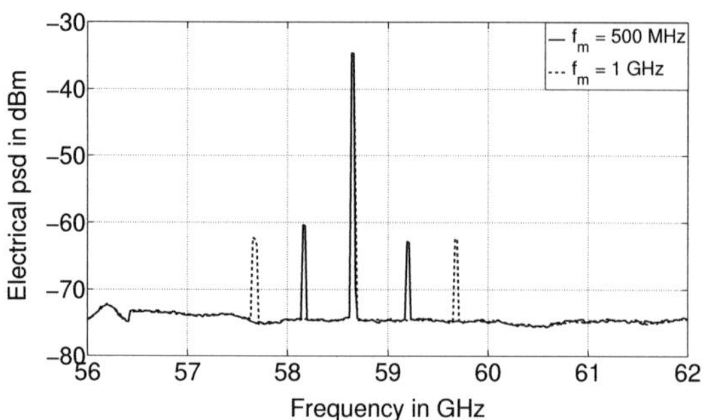

Figure 3.6: Measured asymmetry of the beat spectrum of a directly modulated laser (L34), RBW = 3 MHz.

$\beta = \Delta f / f_m$ is the FM index[13]. The modulated laser spectrum is obtained by a translation of the spectrum $S_m(v)$ of the modulating signal to the laser center frequency v_0, and a convolution of each sideband with the MLLD spectrum. With equation (2.15), we obtain

$$S_{m,opt}(v) = S_m(v) * S_{opt}(v) = S_m(v) * |E_{RF}(v - v_0)|^2 \sum_{|m| \leq \frac{M}{2}} \frac{2\Delta f_m}{(v - f_m)^2 + \Delta f_m^2}.$$

(3.13)

The IM/FM optical spectrum is determined by the IM index m_i and the Bessel functions J_n. For very low IM indices $m_i \ll 1$, or pure FM, the optical spectrum is symmetric about the center frequency. The IM causes the spectrum to become asymmetric. Through the mode-sideband beating, this asymmetry also appears in the electrical beat spectrum. From figure 3.6, it can be observed that in our case, the asymmetry of the beat spectrum varies also with frequency: at a modulation frequency of $f_m = 500$ MHz (solid curve), the lower sideband is clearly more

---

[13] The frequency deviation is determined by I_m and the magnitude of the laser's current to frequency modulation transfer function, evaluated at the bias point I_B and the modulation frequency f_m, a concept used in [OJ82] and references therein.

powerful than the upper sideband. At $f_m = 1$ GHz (dashed curve), the sidebands are rather well balanced. Further single-tone experiments were therefore carried out at $f_m = 1$ GHz, and in the case of two-tone modulation, $f_2 = 1.1$ GHz, where the spectrum is symmetric. The asymmetry of the electrical DSB spectrum is not a problem in itself, as only one of the sidebands will be selected for transmission anyway. However, for broadband modulation, different spectral components in the *same* sideband will have slightly different powers due to this parasitic FM effect and the resulting spectrum might therefore be distorted. In anticipation of the results of the broadband transmission experiments shown in chapter 6, we can nevertheless disregard parasitic FM, as its effects are negligible compared to other imperfections such as frequency jitter or phase noise.

Modulation efficiency

If bond wire parasitics and substrate losses are neglected, the laser equivalent circuit in figure 2.17 effectively consists of an ohmic resistance R_L in series with the active laser medium and a parallel junction capacitance C_L. Thus, for a simple analysis, we can model the laser with this RC circuit which essentially determines its low-pass response. The equivalent circuit is depicted in figure 3.7.

Under small-signal modulation with a modulation signal i_g around a current bias point I_B, the modulating current i_L through the laser active zone is

$$i_L = \frac{1}{j2\pi f C_L R_L + 1} \cdot i_g, \tag{3.14}$$

and the current delivered by a generator with a source voltage v_g and an output impedance R_g into the laser is

$$i_g = v_g \cdot \frac{j2\pi f C_L R_L + 1}{j2\pi f C_L R_L R_g + R_g + R_L}. \tag{3.15}$$

The generator with an available power of $p_{g,a}$ delivers a power of p_g

$$p_g = p_{g,a} \cdot (1 - |\Gamma_L|^2) = \frac{v_g^2}{4R_g} \cdot (1 - |\Gamma_L|^2) \tag{3.16}$$

to the unmatched laser diode, where Γ_L is the reflection factor of the laser diode,

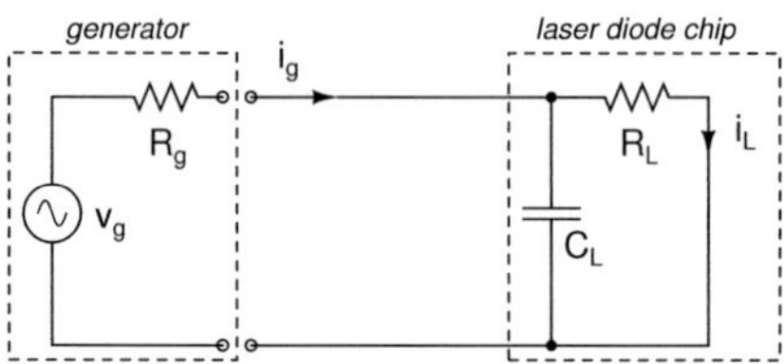

Figure 3.7: Simplified circuit model of the laser diode. The laser is represented by a parallel circuit of R_L and C_L.

$$\Gamma_\mathrm{L} = \frac{(C_\mathrm{L}||R_\mathrm{L}) - R_\mathrm{g}}{(C_\mathrm{L}||R_\mathrm{L}) + R_\mathrm{g}} = \frac{\frac{R_\mathrm{L}}{j2\pi f C_\mathrm{L} R_\mathrm{L} + 1} - R_\mathrm{g}}{\frac{R_\mathrm{L}}{j2\pi f C_\mathrm{L} R_\mathrm{L} + 1} + R_\mathrm{g}}. \tag{3.17}$$

The modulated optical power[14] $p_\mathrm{m,opt}$ is the product of the laser current and the slope efficiency $s_1(i_\mathrm{L} = I_\mathrm{B}) = s_\mathrm{1,B}$:

$$p_\mathrm{m,opt} = s_\mathrm{1,B} \cdot i_\mathrm{L} = s_\mathrm{1,B} \cdot \frac{v_\mathrm{g}}{j2\pi f C_\mathrm{L} R_\mathrm{L} R_\mathrm{g} + R_\mathrm{g} + R_\mathrm{L}}, \tag{3.18}$$

and the frequency-dependent modulation efficiency resulting therefrom is:

$$\frac{p_\mathrm{m,opt}^2}{p_\mathrm{g,a}} = \frac{s_\mathrm{1,B}^2}{(j2\pi f C_\mathrm{L} R_\mathrm{L} R_\mathrm{g} + R_\mathrm{g} + R_\mathrm{L})^2} \cdot \frac{4R_\mathrm{g}}{(1 - |\Gamma_\mathrm{L}|^2)}. \tag{3.19}$$

Equation (3.19) shows implicitly that the inclusion of a parallel capacitance across the laser resistance does not affect the DC or low frequency value of the incremental modulation efficiency, but it rather gives the laser response a low-pass form and thus restricts the bandwidth over which this DC value can be maintained.

The question of how much of the electrical RF power is converted to $p_\mathrm{m,opt}$ - given a certain detector efficiency and optical power - involves two aspects: the frequency-dependent electrical matching of the laser diode to the generator and the efficiency of the E/O conversion, i.e. s_1, which depends on the characteris-

---

[14] its AC contribution

tic PI curve of the laser. The achievable intrinsic link gain is very low for the MLLD-based link, a situation which is detrimental both to the delivered output power of the link and to the link noise figure. We must therefore try to maximize it. In the following sections, techniques to improve incremental modulation efficiency, and thus intrinsic link gain, will be presented. Both the electrical matching and the slope efficiency can be optimized. There is however an important difference between link gain improvements through impedance matching, and through improving modulation efficiency: while the matching most certainly affects the modulation bandwidth of the system, the latter is independent from such limitations.

Impact of slope efficiency

In figure 3.8, the effect of varying slope efficiency on modulation efficiency is shown in simulation. For the unmatched case, i.e. a laser impedance of R_L = 4.5 Ω, and a generator impedance of $R_g = 50\,\Omega$, the maximum modulation efficiency is 0.05 W which can be obtained for a maximum slope efficiency of 0.8 W/A. Resistive matching ($R_L + R_{MATCH} = 50\,\Omega$) actually causes the modulation efficiency to *decrease*, as the power is mostly dissipated in the matching resistor rather than transferred to the laser active zone.

The slope efficiency $s_{l,B}$ is an intrinsic parameter of the MLLD's PI curve. In theory, it should be independent of optical power for currents above the lasing threshold until the laser reaches saturation for high bias currents. As a matter of consequence, the link gain should also be independent from average optical power. In practice, the MLLD's PI curve remains linear only for smaller bias currents, as we have seen in chapter 2, and the choice of bias point becomes critical. In order to maximize link gain, I_B must be chosen in the region of maximum linearity which is situated close to the lasing threshold I_{th} of the characteristic laser curve.

In figure 3.9, the measured link gain is shown for various bias currents, which corresponds to a variation of slope efficiencies, and via equation (3.19) a variation of modulation efficiencies, obtained for sinusoidal modulation. An improvement of more than 9 dB can be obtained in link gain by biasing the MLLD at lower currents. If our *only* goal was to improve link gain and SNR, we could exploit this

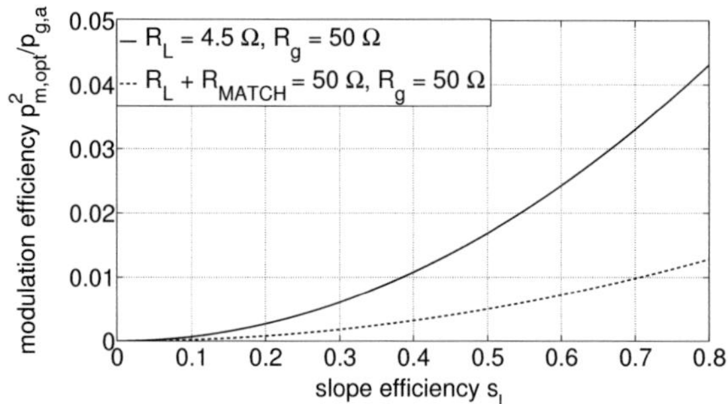

Figure 3.8: Effect of slope efficiency variation on modulation efficiency for a mismatched laser chip modeled by an RC parallel circuit (simulation)

improvement by biasing the laser at a lower injection current. In the absence of other imperfections, a sufficiently high SNR is often synonymous with low error rates. In the presence of frequency or phase instability, however, this frequency and phase noise become the principal limiting factors for transmission quality even at high SNR values. We have seen in chapter 2 that optimum bias points exist with respect to the frequency and phase stability of the generated mm-wave carrier. In our case, a trade-off must be accepted between link gain and stability. In anticipation of the results of the transmission experiments in chapter 6, the SNR values provided by the link are high enough for the link to be limited by frequency jitter and phase noise, rather than by SNR. The optimum bias points must therefore be chosen with respect to stability, although they do not correspond to optimum link gain. These have already been shown in table 2.1.

Slope efficiency can also be improved by optimizing the chip-to-fiber coupling efficiency so that more optical power is coupled into the fiber for a given bias point I_B.

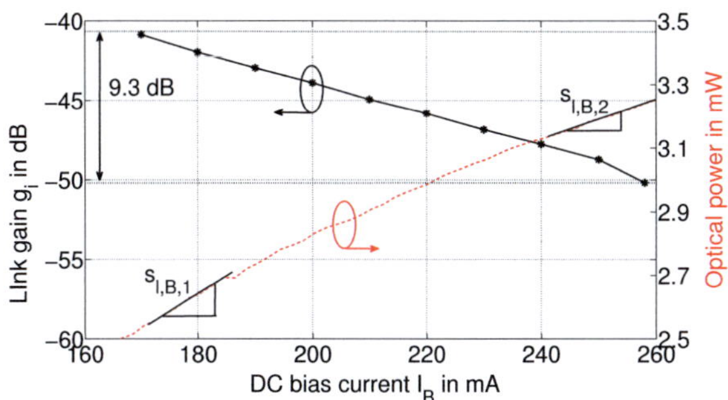

Figure 3.9: Measured variation of p_{load} by varying slope efficiencies $s_{\text{L,B}}$ through the choice of I_B (L34)

Impact of passive matching

As was shown in the previous section, resistive matching does not bring about an improvement of incremental modulation efficiency. In spite of this, it is sometimes done, especially in cases where reflections into the modulation source must be avoided. Distortion effects can be mitigated by including a resistive match [CI04]. A more promising option is that of the reactive match by transmission line stubs or transmission line transformers, as pointed out in section 2.2.4.3. Reactive matching is evidently dependent on frequency. Bode [Bod45] and Fano [Fan50] have derived relationships stating limits to the optimum matching one can obtain across a bandwidth Δf for four generic cases of load impedance where both series and parallel circuits of a capacitor and a resistor, or an inductor and a resistor, are considered [Poz05]. For the parallel circuit of laser resistance R_{L} and C_{L}, we can estimate the lower limit for the magnitude of the reflection factor using

$$|\Gamma_{\text{L}}| = \exp\left(-\frac{1}{2\Delta f \cdot R_{\text{L}} C_{\text{L}}}\right). \tag{3.20}$$

Equation (3.20) is a limit that cannot be realized physically with a finite number

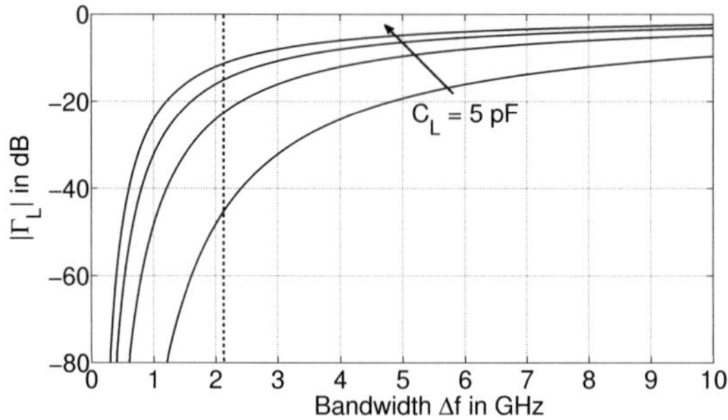

Figure 3.10: Bode-Fano limits for $R_L = 4.5\ \Omega$ and $C_L = 5$ to 20 pF in steps of 5 pF. Dashed line is $\Delta f = 2.16$ GHz

of elements. As such, the limit represents a design target which can only be approximated as closely as possible. When we use equation (3.20), we assume that the substrate losses and bond parasitics of the MLLD package can be neglected and that the matching network is lossless.

In figure 3.10, we can observe the magnitude of the reflection coefficient $|\Gamma_L|$ in dB as a function of the bandwidth Δf, corresponding to the Bode-Fano limits for a laser resistance of $R_L = 4.5\ \Omega$ and a junction capacitance C_L that varies from 5 pF to 20 pF in steps of 5 pF. For a 60 GHz transmission system compliant with the radio standards presented in chapter 1, the bandwidth of one transmission channel is $\Delta f = 2.16$ GHz [IEE09] around an intermediate frequency which ultimately depends on the MLLD beat frequency ν_F in order to give band center frequencies at 58.32 GHz, 60.48 GHz, 62.64 GHz, 64.8 GHz for the radio channels specified in [IEE09]. The relative bandwidth thus is high; for an MLLD with a hypothetical ν_F of 60 GHz, it is 45% minimum. In figure 3.11, the Bode-Fano limit for $\Delta f = 2.16$ GHz is shown as a function of the time constant $R_L C_L$ of the laser package.

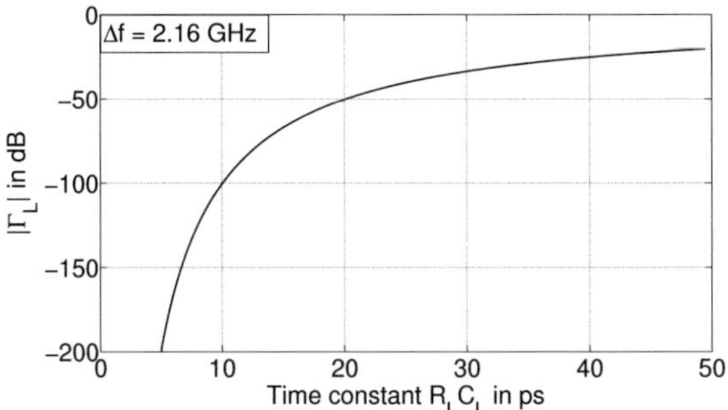

Figure 3.11: Bode-Fano limit for $\Delta f = 2.16$ GHz as a function of the time constant $R_\mathrm{L}C_\mathrm{L}$ of laser resistance and junction capacitance

3.2.1.2 Intrinsic link gain for external modulation

Intensity modulation by means of an MZM has been studied in detail ([CI04], [KSS01]). The principal findings will be mentioned here in the context of the MLLD-based link. In an external modulation scenario, the optical field at the output of the modulator is, with equation (2.4),

$$e_{\mathrm{opt,out}}(t) = \sqrt{T_\mathrm{i}} \cdot e_{\mathrm{opt}}(t) \cdot \exp\left(j\frac{\pi v(t)}{V_\pi}\right)$$
$$= \sqrt{T_\mathrm{i}} \cdot e_{\mathrm{RF}}(t) \cdot \exp(j2\pi v_0 t) \cdot \exp\left(j\frac{\pi v(t)}{V_\pi}\right) \tag{3.21}$$

for a perfectly balanced single-electrode MZM. T_i is the transmission of the modulator from fiber input to fiber output. Under the assumption that the electrode is matched to 50 Ω, the voltage on the electrode is

$$v(t) = V_\mathrm{B} + v_\mathrm{m}(t) = V_\mathrm{B} + V_\mathrm{m}\cos(2\pi f_\mathrm{m}t) = V_\mathrm{B} + \sqrt{8 \cdot p_{\mathrm{g,a}} \cdot R_\mathrm{g}} \cdot \cos(2\pi f_\mathrm{m}t), \tag{3.22}$$

where V_m is the amplitude of the modulation signal and

$$p_{g,a} = \frac{V_m}{8R_g} \qquad (3.23)$$

the available power of the generator. The incremental modulation efficiency for a linearly biased MZM is

$$\frac{p_{m,opt}^2}{p_{g,a}} = \left(\frac{T_i \overline{P} \pi}{2V_\pi} \right)^2 \cdot R_m. \qquad (3.24)$$

The interested reader is referred to [CIABP06] and to appendix B. From equation (3.24) we observe that the efficiency of the modulation depends on T_i and V_π, both of which are native parameters of the modulator[15]. We see that the incremental modulation efficiency of the MZM can be maximized by reducing the half-wave voltage V_π in the design of the device, or by augmenting the value of T_i. However, for any given MZM, V_π and T_i are not parameters accessible to the system designer. A parameter the link designer can however influence is the average optical power at the input $\overline{P}$.

We thus observe that, *in principle*, there is no upper limit to incremental modulation efficiency for the external modulation link. For a given MZM, it should be possible to increase the intrinsic link gain g_i by simply increasing the average optical power at the input of the modulator[16]. An external modulation IF-over fiber link with positive gain in the order of 15 dB has been demonstrated [ABB$^+$05][17]. In practice, the maximum optical power the modulator waveguides can handle is limited.

---

[15] Equation (3.24) is valid only for linear bias. MZM modulation efficiency thus depends also on V_B.

[16] Obtaining positive gain from a cascade of apparently passive components might strike the reader as contradictory. In his comprehensive study of the general performance of direct and external modulation analog fiber optic links, Cox, III gives a plausibility justification: while the available source power drawn by the modulator is independent of the optical power flowing through its waveguides, the modulated power delivered to the load of the photo-detector is clearly dependent on the optical power incident on the detector. In this context, he also discusses the analogy of an MZM and a field effect transistor, where a gate voltage can control a large source-to-drain current (in analogy to CW optical power), even if the gate current itself is negligible [CI04].

[17] Positive link gain of 3.8 dB has also been demonstrated for directly modulated lasers [CIRRH98]. The link relied on a cascade laser architecture.

Table 3.1: Link gain simulation parameters

MLLD parameters	MZM parameters	PD filter effect	Optical losses
$s_{l,B} = 0.0118$	$\overline{P} = 4$ dBm	$T_b = -26.5$ dB	$(T_L)^2_{dB} = -(0.03 \text{ dB} + 1 \text{ dB})$
$R_L = 4.5\ \Omega$	$T_i = 0.25$		*attenuation + connectors*
	$V_\pi = 4.8$ V		
	$R_m = 50\ \Omega$		

For the externally modulated MLLD, we therefore conclude that optimum gain performance is obtained for high optical power at the input of the MZM. Thus, the MLLD is preferably biased at high bias currents for high optical power. Again, we can identify a principal trade-off in biasing the MLLD, namely between maximum optical power and stability. As can be observed from table 2.1, this constraint is less bothersome as the optimum bias points for maximum stability correspond to power values close to saturation.

In terms of intrinsic link gain, raising the average optical power is always favorable. On the other hand, gain is only one of the link parameters we have identified at the beginning of the chapter. Optimum performance implies also a low noise figure. As we will see in section 3.2.2.1, increased optical power usually causes the system noise figure to increase. The trade-off has been studied in [ACIR98].

3.2.1.3 Measurement results for single-tone modulation

Based on equation (3.4), and the respective expressions for direct and external link modulation, the expected link gain can be calculated. The values of the different parameters are listed in table 3.1. In figure 3.12(a), the calculated curves for intrinsic link gain are shown for one of the electrical sidebands. The laser impedance was modeled as purely resistive, as we have seen that the complex laser impedance influences the frequency response, but not the maximum value of the modulation efficiency ("DC" value). Compression effects were not simulated for the direct modulation link for the simple reason that compression was not observed for the intrinsic link and at power values within the range of opera-

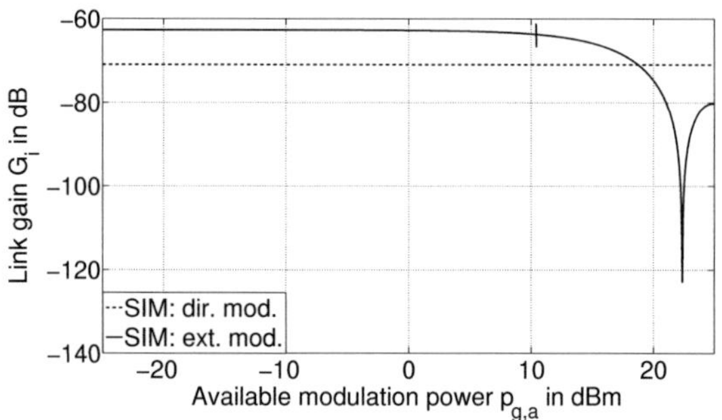

(a) Simulated link gain for one electrical sideband

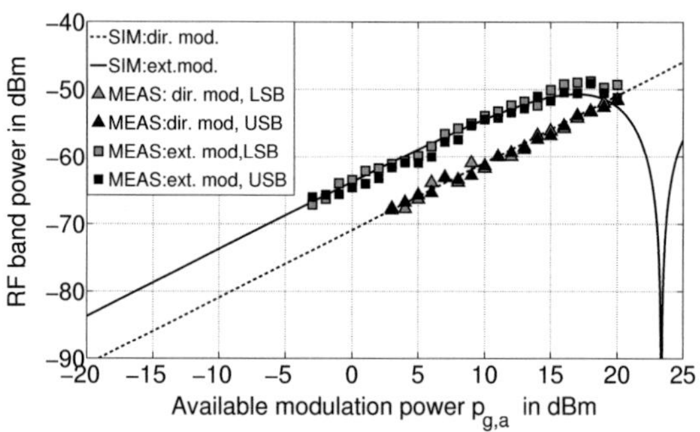

(b) Output power measured in the upper sideband (L34)

Figure 3.12: Output power and link gain simulations and measurements for the direct and the external modulation link.

tion of the MLLD. On the other hand, the external modulation link is in general limited by the compression of the MZM. The gain curve in figure 3.12(a) reflects the fundamental gain compression. We postpone further discussion of link compression until section 3.2.3.

Due to the presence of the optical amplifier, the external modulation link is, strictly speaking, not an intrinsic link anymore. However, since the EDFA was only used to compensate for the optical loss of the MZM, without actually producing optical gain in comparison to the direct modulation link, it will not affect the gain or power performance of the link. Its inclusion will nevertheless change the noise figure of the link, and thus, the available SNR. These effects will be treated in section 3.2.2.

Based on the gain values, the expected output power can be calculated. Figure 3.12(b) shows a comparison of output power measurements and simulations for the electrical upper sideband for both direct (triangles) and external modulation (squares). We observe that the measurements closely follow the predicted values. Gain compression in the external modulation link causes the output power to saturate (section 3.2.3). Throughout the measurements, we always select the upper sideband, even if we have stated before that the sidebands are symmetric and therefore seemingly identical in terms of power. A justification will be given in section 3.2.4.

3.2.2 Link noise figure

In the intrinsic MLLD-based fiber optic link, the overall noise figure results from various noise sources, the dominant ones being the thermal noise in real impedances that appear in the signal path, the RIN of the MLLD itself, and the shot noise in the detector[18]. Following the superposition theorem for linear networks, multiple uncorrelated noise sources can be combined by adding their mean-square voltages or currents, or in statistical terms, the variances of the noise voltages or currents, representing the power delivered to the standard load R_0. The detected noise power of the intrinsic link is then

$$\sigma_{\mathrm{d,i}}^2 = \sigma_{\mathrm{RIN}}^2 + \sigma_{\mathrm{sn}}^2 + \sigma_{\mathrm{th}}^2, \tag{3.25}$$

where we consider laser RIN noise, shot noise and thermal noise in the intrinsic link.

A device's noise figure NF is defined as "the [logarithmic value of the] ratio of the available output noise power per unit bandwidth to the portion of that

---

[18]Shot noise effects in the laser source are usually negligible compared to RIN.

noise caused by the actual source connected to the input terminals of the device, measured at a standard temperature of 290 K" [HAB$^+$59]. While the overall system noise figure can theoretically be determined by cascading the gains and noise figures of the different components, the measurement of the component noise figures is rather impractical. This is why, in this context, we will determine the link noise figure by comparing the noise at the input σ_{in}^2 and the output while at the same time accounting for the link gain g_{i}, where the noise at the output is $\sigma_{\mathrm{d}}^2 = g_{\mathrm{i}} \cdot \sigma_{\mathrm{in}}^2 + \sigma_{\mathrm{add}}^2$. Here, σ_{add} is the amount of noise added by the link:

$$NF_{\mathrm{link}} = 10\log\left(\frac{\sigma_{\mathrm{d}}^2}{g_{\mathrm{i}} \cdot \sigma_{\mathrm{in}}^2}\right) = 10\log\left(1 + \frac{\sigma_{\mathrm{add}}^2}{g_{\mathrm{i}} \cdot \sigma_{\mathrm{in}}^2}\right). \tag{3.26}$$

The measured noise figure NF in decibels is taken as the difference between SNR_{in} and SNR_{out}. For the following considerations, we will closely follow the notation of [ACIB$^+$98].

3.2.2.1 Contributions to output noise

We will examine the different noise sources in the MLLD-based link, calculate the respective noise figures and compare the result to the measured values. We will express noise through its noise current equivalent, or through the corresponding power.

First we consider the source at the input of the link which delivers the modulation signal. At a temperature T, the mean-square[19] thermal noise current from the ohmic part of the source impedance R is

$$\left\langle i_{\mathrm{th}}^2 \right\rangle = \frac{4k_{\mathrm{B}}T\Delta f}{R}. \tag{3.27}$$

For source matching, $R = R_{\mathrm{g}}$. The corresponding input noise power over a bandwidth Δf is then

$$\sigma_{\mathrm{th,in}}^2 = k_{\mathrm{B}}T\Delta f. \tag{3.28}$$

---

[19]The operator $\langle x \rangle$ takes the mean-square value of a quantity.

Δf is the measurement bandwidth of the electrical spectrum analyzer (its resolution bandwidth). At an ambient temperature of $T_0 = 290$ K, thermal noise is flat over a wide range of frequencies[20]. Thus, the thermal noise level is considered constant at $4 \cdot 10^{-21}$ J, or -174 dBm if we assume $\Delta f = 1$ Hz.

We then consider the noise added by the laser. MLLD noise increases with the intensity of the optical signal: laser RIN was introduced in chapter 2 as the common measure for random fluctuations in the laser's optical output power. We have seen that the RIN of the MLLD is strongly dependent on the pump current[21]. At the optimum bias points of about 250 mA for chips L34 and L872, and at the mm-wave operating frequencies, the RIN level can however be considered approximately constant at -160 dBc/Hz. A RIN definition that relates the measured level to the RIN noise current is given by

$$RIN_{\text{MLLD}} = 10\log\left(\frac{2 \cdot \left\langle i_{\text{rin}}^2\right\rangle}{I_{\text{ph,DC}}^2} \cdot \Delta f\right), \tag{3.29}$$

where the factor of 2 accounts for the integration of the two-sided frequency spectrum and $I_{\text{ph,DC}}$ is the average DC photo-current from equation (2.23). After rearranging, the equivalent mean-square noise current for RIN is

$$\left\langle i_{\text{rin}}^2\right\rangle = \frac{I_{\text{ph,DC}}^2}{2} \cdot 10^{\frac{RIN_{\text{MLLD}}}{10}} \cdot \Delta f, \tag{3.30}$$

and the RIN noise power delivered to a load R_0 is

$$\sigma_{\text{RIN}}^2 = \left\langle i_{\text{rin}}^2\right\rangle \cdot R_0. \tag{3.31}$$

Finally, shot noise is the limiting noise type on the detector's side. This particular noise arises in a situation where a current - here the photocurrent - is established due to the quantized transits of random charge carriers [Sch18], as in the case of a semiconductor diode. The shot noise current is superimposed on the average

---

[20]H. Nyquist showed that thermal noise is flat over a range of frequencies limited by $0 \leq f \ll k_{\text{B}}T/h$ [Nyq28].

[21]It is furthermore weakly dependent on frequency (see figure 2.15(b)) which we will neglect.

DC photocurrent. Its mean-square value is:

$$\langle i_{sn}^2 \rangle = 2q \cdot I_{ph,DC} \cdot \Delta f, \tag{3.32}$$

and the corresponding noise power delivered to R_0 is

$$\sigma_{sn}^2 = \langle i_{sn}^2 \rangle \cdot R_0. \tag{3.33}$$

All noise powers superimpose at the photo-detector to give σ_d^2 according to equation (3.25).

3.2.2.2 Direct modulation link with and without electrical amplification

In the direct modulation MLLD-based link, the noise added by the link results from the thermal noise of the laser (or its real impedance), amplified by the link gain, as well as the thermal noise of the detector real impedance, MLLD RIN, and detector shot noise,

$$\sigma_{add,direct}^2 = k_B T \Delta f \cdot g_i + k_B T \Delta f + (\langle i_{rin}^2 \rangle + \langle i_{sn}^2 \rangle) \cdot R_0. \tag{3.34}$$

With equation 3.26, the intrinsic link noise figure is then

$$NF_i = 10\log\left(2 + \frac{1}{g_i} + \frac{(\langle i_{rin}^2 \rangle + \langle i_{sn}^2 \rangle) \cdot R_0}{k_B T \Delta f g_i}\right). \tag{3.35}$$

When, as in the case of the MLLD, the intrinsic link gain is very low, the noise figure approaches a value equal to the RF link loss, and both shot noise and RIN noise are negligible compared to g_i. The lower limit for link noise figure is thus [ACIR98]:

$$\lim_{\langle i_{rin}^2 \rangle + \langle i_{sn}^2 \rangle \to 0, g_i \to 0} NF_i = 10\log\left(2 + \frac{1}{g_i}\right), \tag{3.36}$$

known in the literature as the *passive attenuation limit* [CI04]. For negative link gain (i.e. gain values $g_i < 1$), the noise figure increases in reciprocal proportion to intrinsic link gain. In figure 3.13, we have plotted the measured noise figure value for the direct modulation MLLD link (white triangle), as well as the passive attenuation limit. The intrinsic direct modulation MLLD link operates very close to this limit. In order to improve the link noise figure, the measures proposed in the previous section for improving link gain can be applied.

With an intrinsic link gain of about $G_i = $ -65 dB (see table 3.2), the MLLD-based direct modulation link cannot provide sufficiently high output power at the photo-detector. An electrical amplifier with a linear noise factor of F_{el} can be employed to boost the electrical power at the BS. Its inclusion will affect the link noise figure: The total noise figure can be determined using the Friis equation [Fri44], where the noise factor F_i is the linear representation of NF_i,

$$NF_{total} = 10\log\left(F_i + \frac{(F_{el}-1)}{g_i}\right).$$ (3.37)

The noise figure limit for a link where shot noise and RIN noise become negligible compared to the large negative intrinsic gain is then

$$\lim_{\langle i^2_{rin}\rangle + \langle i^2_{sn}\rangle \to 0, g_i \to 0} NF_{total} = 10\log\left(2 + \frac{F_{el}}{g_i}\right).$$ (3.38)

The link is not passive anymore, but it is still gain-limited in the sense that the reciprocal link gain is the determining factor in calculating the total noise figure. The curve describing the relationship of NF_{total} on G_i is plotted again in figure 3.13 (dashed line, "gain-limited"), as well as the measured value (grey triangle). We thus observe that including an electrical RF amplifier might be necessary to raise the signal above the sensitivity of the receiver, but in terms of noise figure and SNR, we must expect further degradation.

3.2.2.3 External modulation link with electrical and optical amplification

In the case of the external modulation link, we find again the same noise sources as in the direct modulation link. Thermal noise in the link now originates from the real impedance of the MZM electrode. The MLLD-based external modula-

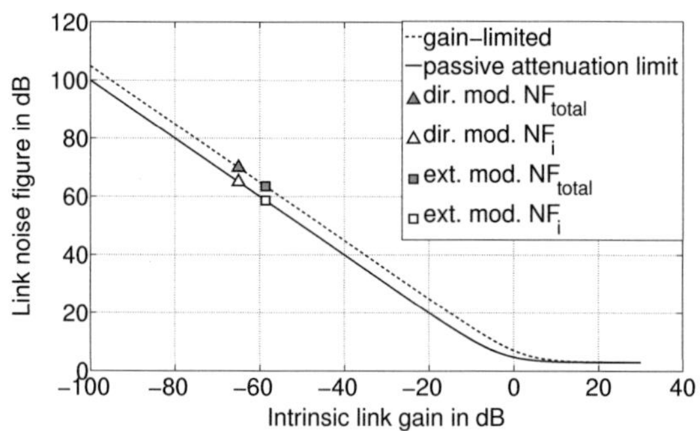

Figure 3.13: Link noise figures in the gain-limited link

tion link contains the same MLLD RIN level as before, and on the detector side, we again expect to find the effects of shot noise discussed in the previous section. Equations (3.35) and (3.37) therefore remain valid and only need to be completed in the cases where an amplifier is also included in the optical domain.

The use of an optical amplifier, e.g. an EDFA, is inevitable when a less powerful diode (such as e.g. L34, centered around 1550 nm) is used and the insertion loss of the MZM needs to be compensated for. It has been shown that the inclusion of an EDFA can enhance the electrical SNR[22] at the detector [Des94], which is why it is typically used as a pre-amplifier for photo-receivers in medium- to long-distance fiber links. The output of the EDFA is usually filtered to cut off a large part of its noise spectrum. As we do not filter EDFA noise from the output signal of the EDFA, we cannot expect such improvement, but even have to accept an increase in output noise power.

In the MLLD-based RoF link, the introduction of the EDFA implies a modification of the MLLD spectrum according to the EDFA's gain spectrum $G_e(v)$. The EDFA also represents an additional noise source.

---

[22] At first glance, this result might seem contradictory, as an EDFA always exhibits an *optical* noise figure ≥ 3 dB. Desuvire shows however that such improvement is possible in the *electrical* domain [Des94].

The mode field amplitude introduced in section 2.1.1.1 must be recalculated:

$$\left|A'_{\mathrm{m}}\right| = \sqrt{G_{\mathrm{e}}(v_{\mathrm{m}})} \cdot \left|A_{\mathrm{m}}\right|. \tag{3.39}$$

With equations (2.21) - (2.28) the resulting power p'_{RF} of the received signals can be determined, affecting the numerator in the SNR value.

Naturally, the amplifier introduces noise (represented by its noise figure), mainly due to amplified spontaneous emission (ASE) in its transmission band. Theoretically, the noise figure of the EDFA is limited to 3 dB [Des94]. Practically, it depends strongly on the drive level of the amplifier. Rewriting equation (3.25), the total detected noise of the link now contains a contribution of the EDFA,

$$\sigma_{\mathrm{d}}'^{,2} = \sigma_{\mathrm{RIN,amp}}^2 + \sigma_{\mathrm{sn,amp}}^2 + \sigma_{\mathrm{sig-ASE}}^2 + \sigma_{\mathrm{ASE-ASE}}^2 + \sigma_{\mathrm{th}}^2, \tag{3.40}$$

where two additional noise powers $\sigma_{\mathrm{sig-ASE}}^2$ and $\sigma_{\mathrm{ASE-ASE}}^2$ have been included, and the RIN noise term as well as the shot noise term need to be modified in order to account for the amplification of the optical signal:

$$\sigma_{\mathrm{RIN,amp}}^2 = G_{\mathrm{e}}^2 \cdot \sigma_{\mathrm{RIN}}^2, \tag{3.41}$$

$$\sigma_{\mathrm{sn,amp}}^2 = G_{\mathrm{e}} \cdot \sigma_{\mathrm{sn}}^2. \tag{3.42}$$

Note that the RIN noise is proportional to the square of the amplifier gain, because the photocurrent depends linearly on the optical power and contributes quadratically to output power, whereas shot noise depends linearly on the photocurrent. $\sigma_{\mathrm{sig-ASE}}^2$ and $\sigma_{\mathrm{ASE-ASE}}^2$ refer to noise generated by the beating between the signal and the amplifier's ASE noise, and to the beating of ASE with itself.

Signal-ASE beat noise can be determined as follows:

$$\sigma_{\mathrm{sig-ASE}}^2 = 4G_{\mathrm{e}} \langle i_{\mathrm{ase}} \rangle R_0 I_{\mathrm{ph,DC}} \frac{\Delta f}{B_{\mathrm{o}}}, \tag{3.43}$$

where B_{o} and Δf are the optical and electrical bandwidths considered. If no filtering is applied in the optical domain, B_{o} covers the entire range of the EDFA gain spectrum in the optical C-band (1530 nm to 1565 nm). Finally, $\langle i_{\text{ase}} \rangle$ is the mean ASE noise current, and it can be shown that it is calculated as [Des94]:

$$\langle i_{\text{ase}} \rangle = \rho q \sigma_{\text{eq}}^2 G_{\text{e}} B_{\text{o}}, \tag{3.44}$$

where σ_{eq}^2 is the equivalent input noise of the EDFA.

The last contribution to detector noise is ASE-ASE beat noise. It is calculated as

$$\sigma_{\text{ASE-ASE}}^2 = M_{\text{p}} \left\langle i_{\text{ase}}^2 \right\rangle R_0 \frac{2\Delta f}{B_{\text{o}}} \propto B_{\text{o}}. \tag{3.45}$$

M_{p} is the number of polarization modes. In SMF, it is $M_{\text{p}} = 2$. For the noise power calculations in equations (3.43) and (3.45), we have assumed that the ASE spectrum emulates the gain spectrum of the EDFA, and that the gain spectrum $G_{\text{e}}(v) = G_{\text{e}}$ is flat over B_{o}. These assumptions are valid only for sufficiently high input powers. As can be observed from figure 3.14(a), the noise floor then tends to flatten, especially with regard to the characteristic bump around 1530 nm.

The noise figures for the measurement with and without the electrical amplifier are shown in figure 3.13 (squares). When the noise figure for the external modulation link was measured, an EDFA was included in the setup. Both values fall on the curves for the passive attenuation link and the gain-limited link, respectively. We can therefore conclude that the influence of ASE noise on link noise figure is negligible, as is that of RIN, shot or thermal noise, due to the high negative link gain.

Another effect should be mentioned at this point. A mismatch between the operating bandwidth and the wavelength of the input signal causes strongly detrimental effects on the signal quality, such as spectral distortion and highly increased noise level, which is why the EDFA cannot be used in combination with mode-locked laser diode L872 (centered around 1570 nm). The effects of amplifier distortion due to out-of-band operation are shown in figure 3.14(b).

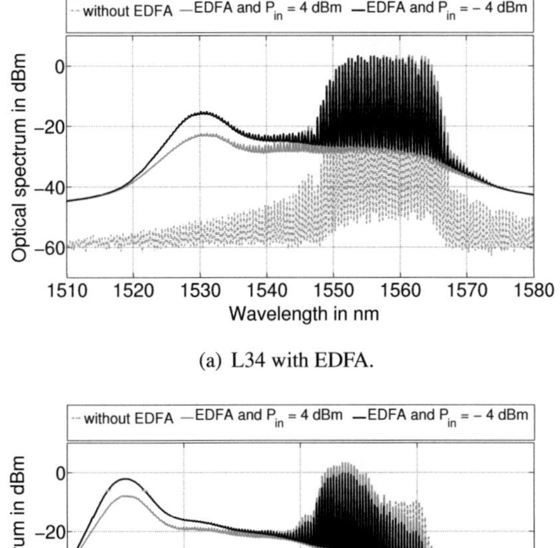

(a) L34 with EDFA.

(b) L872 with EDFA

Figure 3.14: OSA measurements: Spectral distortion and noise contribution of the EDFA. Resolution 0.07 nm.

3.2.2.4 Optimizing signal-to-noise ratio by applying optical filtering in an EDFA-amplified link

Even for the passband amplification of a (single-mode) laser signal, the level of ASE is usually considered unacceptably high, and a narrow band-pass filter is used in order to suppress the noise around the amplified signal. Although the

MLLD spectrum takes up a considerable width in the EDFA passband, the MLLD signal should ideally also be filtered to lower ASE noise power on the detector. From figure 3.14(a), we see that the EDFA used in combination with L34 allows for a gain of about 20 dB, but at the same time its inclusion raises the noise floor considerably. Filtering could not be implemented as appropriate filters, e.g. fiber Bragg gratings, were not available[23]. The question as to how much could have been gained in terms of noise figure and SNR when filtering the MLLD spectrum shall nevertheless be discussed.

We assume that we can filter the entire MLLD spectrum and thus cut-off as much ASE noise as possible without limiting the MLLD spectrum. The noise level at the output of the detector is reduced while the signal power is maintained. In order to determine the relevant noise or signal power terms, equation (3.44) is inserted into equation (3.43). Doing so reveals that as long as no part of the spectrum contributing to signal power is filtered the signal-ASE beat noise is independent of the optical bandwidth B_o and can therefore not be improved by an optical filter device. The only relevant term is then ASE-ASE beat noise. From equation (3.45), we can observe that ASE-ASE beat noise is proportional to B_o and can be reduced by filtering. We can estimate the SNR enhancement x_I which could be gained in the best case from filtering the MLLD spectrum at the output of the EDFA, assuming for simplicity an ideal and lossless filter:

$$
\begin{aligned}
x_{\mathrm{I,dB}} &= SNR'_{\mathrm{dB}} - SNR_{\mathrm{dB}} \\
&= 10\log\left(\frac{\sigma_{\mathrm{d}}^2}{\sigma_{\mathrm{d}}'^2}\right) \\
&= 10\log\left(\frac{\sigma_{\mathrm{RIN,amp}}^2 + \sigma_{\mathrm{sn,amp}}^2 + \sigma_{\mathrm{sig-ASE}}^2 + \sigma_{\mathrm{th}}^2 + \sigma_{\mathrm{ASE-ASE}}^2}{\sigma_{\mathrm{RIN,amp}}^2 + \sigma_{\mathrm{sn,amp}}^2 + \sigma_{\mathrm{sig-ASE}}^2 + \sigma_{\mathrm{th}}^2 + k_o \cdot \sigma_{\mathrm{ASE-ASE}}^2}\right),
\end{aligned}
\tag{3.46}
$$

where $k_o = B'_o/B_o$ is the fractional bandwidth, or the ratio between the optical EDFA bandwidth B_o and the optical filter bandwidth B'_o. As can be observed, the

---

[23] An appropriate filter should have allowed to filter two modes with a spacing of about 60 GHz at a time.

SNR enhancement depends on the absolute level of aggregate noise power. If we want to recover the full MLLD spectrum, yet filter undesired noise, the fractional bandwidth must lie somewhere between $1/2 \cdot B_\mathrm{o}$ and $1/3 \cdot B_o$ (see figure 3.14(a)). The absolute SNR enhancement depends on the ratios of all five noise powers. Assuming for the sake of the argument that all noise powers contribute equally to output noise, the SNR enhancement lies around 0.45 dB ($1/2 \cdot B_\mathrm{o}$) and 0.63 dB ($1/3 \cdot B_\mathrm{o}$), respectively.

On the other hand, we have stated in section 3.2.1 that it is beneficial with respect to both link gain and to dispersion performance (to be discussed in chapter 4) to limit the spectrum to a small number of powerful modes, ideally $M = 2$, which can subsequently be modulated. If the hardware effort is increased by including a filter, we might as well put the EDFA and the filter before the modulator and thus limit the MLLD spectrum. As the link response is simply the product of the response of the different components, the order is arbitrary. We therefore not only reduce the noise level but also improve signal power, as we circumvent the PD low-pass filter effect and achieve a maximization of T_b.

If an arbitrary number of modes of the multimode MLLD spectrum are filtered, everything except the thermal noise power changes, because the signal current and signal power change. The SNR enhancement $x_{\mathrm{II,dB}} = 10 \log \left(\sigma_\mathrm{d}^2 / \sigma_\mathrm{d}''^2 \right)$ cannot simply be represented as a function of the fractional bandwidth k_o. The output noise $\sigma_\mathrm{d}''^2$ is modified as the sum of the following contributions:

$$\sigma_{\mathrm{RIN,amp}}''^2 \propto \left(\sum_{m=1}^{M} |A_\mathrm{m}'|^2 \right)^2 ,$$

$$\sigma_{\mathrm{sn,amp}}''^2 \propto \sum_{m=1}^{M} |A_\mathrm{m}'|^2 ,$$

$$\sigma_{\mathrm{sig-ASE}}''^2 \propto \sum_{m=1}^{M} |A_\mathrm{m}'|^2 ,$$

$$\sigma_{\mathrm{ASE-ASE}}''^2 \propto \frac{1}{B_\mathrm{o}} ,$$

where the amplitudes $|A'_\mathrm{m}|$ correspond to the amplified spectrum according to equation (3.39). At the same time, filtering limits the number of modes and the output signal power. With equations (3.4) and (3.9), we obtain the dependence of p''_out on the number of modes:

$$p''_\mathrm{out} \propto g_\mathrm{i} \propto T_\mathrm{b}.$$

Filtering only two modes from the spectrum ($M = 60 \rightarrow M' = 2$), and assuming again that all noise powers contribute equally to output noise, the achievable SNR enhancement can be as much as 30 dB which is first due to the enhancement of T_b, and second, to the scaling of RIN noise, shot noise and signal-ASE beat noise with the effective reduction of average photocurrent.

3.2.3 Linearity and dynamic range

While the concept of noise always implies a certain amount of randomness and is hence most accurately described by a statistical representation, signals caused by distortion are deterministic in nature. They are attributed to a non-linearity at a precise location in the link[24]. For a device with a non-linear transfer function, such as the MZM or an electrical amplifier, we can expect to see distortion effects on the fundamental tone, commonly referred to as compression. Link gain compression is a direct consequence of the composite non-linearities present in the components. A measure of linearity is the *1 dB compression point*, referred to input or output power (IP1dB, or OP1dB, respectively), the power value for which the gain is compressed by 1 dB.

Distortion is commonly quantified in terms of second or third order intermodulation[25] (SOI, TOI). SOI or TOI are usually quantified by the *second (or third) order intercept point* (IP2, IP3), the point where the extrapolated linear curve fits of the power transfer curves of the fundamental signals and the second (or third)

---

[24] While *harmonic* distortion refers to the creation of new signals at higher harmonics of a single tone in a non-linearity, at least two tones must be present for the non-linearity to produce *intermodulation* distortion at frequencies representing a linear combination of the tones at the input.

[25] A common method to evaluate a device's or a system's intermodulation distortion characteristics is the two-tone test, where two frequencies f_1 and f_2 are applied at the input. For the two-tone test, the modulation signal at the source is $v_\mathrm{m}(t) = V_1 \sin(2\pi f_1 t) + V_2 \sin(2\pi f_2 t)$.

order intermodulation products intersect[26]. They can be referred to either input or output power (IIP2, IIP3, or OIP2, OIP3, respectively). For heterodyne reception, TOI is considered the more severe case of distortion because its products lie closely around the fundamentals and cannot usually be filtered. However, for very large bandwidth systems, the SOI products - separated by an octave of the fundamentals - also fall inside the passband, and SOI distortion represents a real limit to dynamic range in multi gigabit systems that have very wide bandwidth. For the considerations below, we will assume a system which covers at maximum one octave.

When noise and distortion effects are considered, it is common practice to quote the *spurious-free dynamic range* (SFDR). SFDR-2 and SFDR-3 are measures for the total usable range of powers for a given noise level and a given measurement bandwidth[27], limited by SOI and TOI. It is limited at its lower end by the input power of the fundamental tones where the output signal can be distinguished from noise, and at its upper end by the input power that causes the SOI or TOI products to rise above noise level.

Distortion can arise both in the modulation and in the detection devices, as well as in active devices along the link. Insofar as the MLLD-based link is concerned, we will be interested in the level of distortion as present in the electrical spectrum at the signal detection stage, i.e after photo-detection and after electrical amplification. As for noise figure, it is in principal possible to measure the distortion of each component in the link and combine the measured values according to the cascade in order to derive the distortion levels at the output of the system [Ega03]. We must assume that distortion products add perfectly coherently *or* perfectly randomly, and we must take the frequency dependence of several

---

[26] In the context of multi-carrier systems, such as CATV, certain carrier pairs produce intermodulation signals at the same frequencies, although the information transported by the carrier is not identical. Therefore, the power in an intermodulation term is determined considering the semi-coherent addition of multiple intermodulation terms [CI04]. The distortion measures used are termed *composite second order* (CSO: $v_i \pm v_j$, $2v_i$, $2v_j$) and *composite triple beat* (CTB: $2v_i \pm v_j$, $2v_j \pm v_i$, $3v_i$, $3v_j$) tones [Des94]. CSO and CTB will not be considered here.

[27] The dependence on bandwidth is related to the fact that the noise measurement itself is dependent on bandwidth, typically the resolution bandwidth of the spectrum analyzer. A 1-dB increase in noise power causes a net decrease in SFDR-2 (SFDR-3) of 1/2 (2/3), which is why SFDR-2 (SFDR-3) scales as $(\text{bandwidth})^{1/2}$ ($(\text{bandwidth})^{2/3}$), and the corresponding unit is $\mathrm{dB \cdot Hz}^{1/2}$ ($\mathrm{dB \cdot Hz}^{2/3}$).

stages into account[28]. In the MLLD-based link, where we deal with different stages of frequency translation, this would mean basing our calculations on a set of hypotheses which we have no way of validating. This is why we will chose a different approach here and limit our investigation to measuring link distortion and finding a meaningful interpretation of the results. The two devices determining the distortion characteristics in the direct and the external modulation link are the MLLD and the MZM, respectively. Their influence will be discussed in the following.

3.2.3.1 Distortion induced by the mode-locked laser diode in the direct modulation link

Laser distortion typically originates from so-called *clipping* effects, where the combination of modulation and bias currents results in a modulation signal driving the laser below the lasing threshold or well into the saturated region when biased in the linear region. Clipping principally limits the direct modulation performance of a laser diode. Laser distortion depends on the bias injection current into the laser: close to threshold, TOI is minimized, but SOI has a maximum [CI04]. While such general ideas can be gained from the laser's PI curve, it is impractical to draw conclusions on the magnitude of the distortion terms based on a *measured* PI curve - which is all we have for the MLLD. Attempts have been made to derive the magnitude of the modulation terms and the corresponding intercept points from the laser gain characteristics [DB90]. Assumptions must then be made regarding the frequency dependence of the gain medium, and the ratio of the modulation and the relaxation frequencies. As, on the one hand, we lack a closed form solution for the laser, and on the other, we cannot be sure about the validity of the assumptions made in [DB90] in the case of the MLLD, we cannot base our considerations thereupon.

Furthermore, it is usually assumed that the injection of the modulation signal does not perturb the laser operation itself. For the MLLD, the situation is different: a modulation signal injected into the passive cavity might destabilize the laser such that the modes lose coherence. The upper limit to modulation power,

---

[28]The models for component distortion and total distortion in a cascade can accommodate frequency dependence when filters are included [Ega03].

and thus to dynamic range (DR), in the direct modulation link is therefore *a priori* not given by distortion, but by the mode-locking condition itself. Although the phenomenon of mode-locking in the quantum dash laser has not yet been fully understood, this one is a condition we can easily verify by measurement.

3.2.3.2 Distortion induced by the Mach-Zehnder modulator in the external modulation link

Distortion induced by the MZM has been studied early on in the modulator's history [BB84]. It could be shown that even-order distortion is a strong function of modulator bias while odd-order distortion also depends on bias, but in a much weaker form. Biasing the MZM in quadrature for linear operation results in zero second-order distortion. The linear dynamic range is therefore not limited by the second harmonic, but by the third, separated by two octaves from the fundamental. For linear bias at the quadrature point, the MZM can be considered to be limited by TOI only. Unlike for the MLLD, there exists an analytical form for the Mach-Zehnder transfer function (see appendix B). From the transfer function, the magnitudes of the intermodulation products can be derived for two-tone modulation. This standard case has been studied by Kolner and Dolfi [KD87]. Only the main results will be stated here. For a two-tone modulated signal, E/O converted by the MZM and O/E converted by an ideal photo-detector, the photocurrent is

$$i_{\mathrm{ph,mzm}} = \frac{\rho \overline{P}}{2} \left(1 + \cos \left(\frac{\pi V_{\mathrm{B}}}{V_\pi} + \frac{\pi v_{\mathrm{m}}(t)}{V_\pi} \right) \right), \qquad (3.47)$$

where the cosine term is responsible for the distortion of the signals. The MZM is assumed to be biased at the quadrature point, i.e. $\pi V_{\mathrm{B}}/V_\pi = \pi/2$. For the two-tone test, the voltage on the electrode is $v_{\mathrm{m}}(t) = V_{2\mathrm{T}} \cos\left(2\pi f_1\right) + V_{2\mathrm{T}} \cos\left(2\pi f_2\right)$. Using a Bessel function expansion, the term becomes

$$\cos\left(\frac{\pi V_{\mathrm{B}}}{V_{\pi}}+\frac{\pi v_{\mathrm{m}}(t)}{V_{\pi}}\right)=\Re\left\{\exp\left(\frac{\pi V_{\mathrm{B}}}{V_{\pi}}\right)\sum_{k}\sum_{l}J_{k}\left(\frac{\pi V_{2\mathrm{T}}}{V_{\pi}}\right)J_{1}\left(\frac{\pi V_{2\mathrm{T}}}{V_{\pi}}\right)\right.$$
$$\left.\cdot\exp\left(j2\pi(kf_{1}+lf_{2})t\right)\right\}, \tag{3.48}$$

and the order of the intermodulation products is $|k|+|l|$. The distortion of the fundamental tones is governed by

$$J_{0}\left(\frac{\pi V_{2\mathrm{T}}}{V_{\pi}}\right)\cdot J_{1}\left(\frac{\pi V_{2\mathrm{T}}}{V_{\pi}}\right).$$

In the absence of the second tone $v_{\mathrm{m}}(t)=V_{1\mathrm{T}}\cos\left(2\pi f_{1}\right)$, the single tone is compressed according to $J_{0}(0)J_{1}(x)=J_{1}(x)$. This is precisely what we have already observed for single-tone modulation in the link gain or output power measurements without electrical amplification presented in figures 3.12(a) and 3.12(b).

The 1 dB compression point depends on the level of bias. At the quadrature point, the relationship between the half-wave voltage and the voltage on the electrodes leading to a compression of the output power by 1 dB, $V_{1\mathrm{T},1\mathrm{dB}}$ is given by

$$V_{1\mathrm{T},1\mathrm{dB}}=\frac{V_{\pi}}{\pi}, \tag{3.49}$$

and the output power at the OP1dB is

$$P_{\mathrm{out},1\mathrm{dB}}=R_{0}\cdot\left(\frac{\rho\overline{P}}{2}\cdot 2\cos\left(\frac{\pi V_{\mathrm{B}}}{V_{\pi}}\right)\cdot J_{1}\left(\frac{\pi V_{1\mathrm{T},1\mathrm{dB}}}{V_{\pi}}\right)\right)^{2}$$
$$=R_{0}\cdot\left(\frac{\rho\overline{P}}{2}\cdot 2\cos\left(\frac{\pi V_{\mathrm{B}}}{V_{\pi}}\right)\right)^{2}. \tag{3.50}$$

The relevant coefficient for TOI at $f = 2f_2 \pm f_1$ is

$$J_1\left(\frac{\pi V_{2T}}{V_\pi}\right) \cdot J_2\left(\frac{\pi V_{2T}}{V_\pi}\right).$$

Kolner and Dolfi have also derived a relationship for the voltage $v_{TOI} = v_{2T}$ on the electrode that corresponds to the IIP3 power, which is:

$$v_{TOI} = 2\sqrt{2}\frac{V_\pi}{\pi}. \tag{3.51}$$

In the case of the MZM used for the measurements (single electrode impedance of 50 Ω and $V_\pi = 4.8$ V), we would expect an MZM-induced IIP3 of about 33.38 dBm. For two reasons, this value only represents an upper limit to the external modulation link IP3. First, we have assumed an ideal and perfectly matched MZM, as well as an ideal and perfectly matched photo-detector. Second, in the link, intermodulation products originating in other components, such as the electrical amplifier, might add to the MZM intermodulation products. As a consequence, the IP3 level decreases.

3.2.3.3 Distortion induced by other components

Other than by the modulation devices, distortion might also be introduced by other components, such as the photo-detector, the optical and electrical amplifiers or by the mixer in the down-conversion stage.

In practice, distortion induced by higher order non-linear effects in the photo-detector are negligible compared to distortion at the modulation device [CI04]. For the sake of completeness, it should be mentioned that the two principal parameters that the link designer disposes of in order to reduce the distortion of the photo-detector are the incident optical power on the PD and the reverse bias applied to the detector. While distortion usually decreases with increasing reverse bias, its level typically rises with increasing detector current [WED96]. In the transmission experiments shown herein, a commercial photo-detector with fixed reverse bias is used. The bias voltage therefore is not a parameter accessible for optimization. In the direct modulation link, the optical power in the link is determined by the MLLD under test; it does not seem reasonable to lower it further. For the external modulation link, we have seen that it is favorable to modulation efficiency to operate at high optical power. In short, the PD is sufficiently linear as it is - and so a reduction of the available power is not justified.

Non-linearities in electrical amplifiers are a notorious source of link distortion. The mm-wave amplifier used for the experiment has a flat gain of 24 dB across its operating range from 57.0 GHz to 66.0 GHz (including the sideband center frequency $v_F + f_m = 59.63$ GHz for chip L34), and a noise figure of 4.9 dB. It exhibits the following characteristics referred to output RF power: OP1dB = 8.7 dBm (IP1dB = -14.3 dBm), as measured, and OIP3 = 32 dBm. Filtering of the mm-wave band before the amplifier stage is not performed. Even for the low output powers delivered by the intrinsic link ($p_{out} = -56.4$ dBm), distortion effects can be observed for high input powers into the electrical link due to the presence of the relatively high power carrier signals. Thus, the electrically amplified link will possibly suffer from (SOI or TOI) distortion.

Likewise, we expect to see the combined effects of MZM distortion and amplifier distortion in the electrically amplified external modulation link.

In contrast to electrical amplifiers, which are inevitably driven into saturation for high input powers and give rise to signal distortion, the EDFA is very low in distortion[29]. Non-linearity arising from gain saturation is negligible [Des94]. In fact, the EDFA should actually be operated at high input power levels and saturated gain for minimum ASE.

Finally, the down-conversion mixer from figure 3.4 potentially introduces distortion. Yet we can exclude it from our considerations, for compression-free operation is guaranteed by the manufacturer for $p_{LO} + p_{RF} < 19$ dBm which largely suffices for our system (LO driver power is 12 dBm).

---

[29] In contrast to an electrical amplifier in saturation which clips off the input signal and thus produces distortion, i.e. spectral components at new frequencies, an optical amplifier in saturation simply applies less gain to its input regardless of the input signal level, and no signals at new wavelengths are created. A common distortion effect in EDFAs is however the combined effect of laser chirp and the non-uniformity of the amplifier's gain, expressed by the gain tilt $dG(v)/dv$.

3.2.3.4 Measurement results for two-tone modulation

Two-tone tests were performed for four different configurations for chip L34: the intrinsic direct modulation link, the electrically amplified direct modulation link, the optically amplified external modulation link and the same external modulation link including additional electrical amplification. Note that the intrinsic external modulation link was not implemented due to the limited optical power available from L34. Figures 3.15 and 3.16 show the power transfer curves for the different configurations. A summary of the results is presented in table 3.2, where the values printed in bold letters refer to actual measurements, and the others to values derived from the measurements. All measurements were performed on an ESA with a noise floor limitation of -150 dBm/Hz. Referred to the measurement bandwidth (RBW = 100 kHz), this translates to a noise floor of -100 dBm/100 kHz, the floor reference for the power curves shown in figures 3.15 and 3.16.

Intrinsic direct modulation link: from figure 3.15(a), we observe that the maximum input power[30] for stable mode-locking is roughly 19 dBm, and the DR resulting from this is 41 dB in 100 kHz bandwidth. While we might improve the power efficiency of the link by improving the matching condition at the input of the system, this measurement indicates that the obtainable DR will most likely not be increased. The noise level rises to -94.6 dBm/100 kHz; we are therefore not limited by the sensitivity of the ESA. Compression effects could not be observed which is why we have no means of determining the IP1dB/OP1dB for the intrinsic direct modulation link. As we have seen in figure 3.12(b) (section 3.2.1.3), the output power is limited to roughly -56 dBm, resulting in a total link gain of -71.4 dB. In order to calculate the intrinsic link gain, the conversion loss of the down-converter and the combined cable loss must be taken into account, so that the intrinsic link gain is -65 dB.

Electrically amplified direct modulation link: The electrical amplifier boosts the output power to about -32 dBm, so that a total link gain of -47 dB is achieved, while intrinsic link gain, by definition, stays the same. The measured SNR is lowered by 2.1 dB. From the datasheet noise figure, we would have accepted a

---

[30]This power is actually *transferred* to the MLLD; it is $\neq p_{g,a}$ due to the impedance mismatch at the laser.

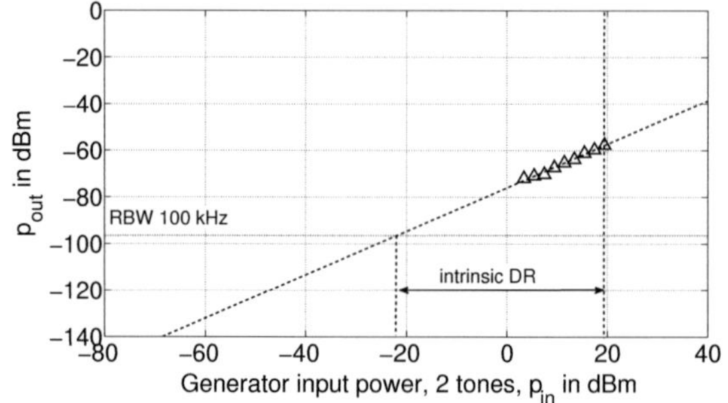

(a) Intrinsic link, measured in the upper sideband. Two-tone signal: f_1 = 1 GHz, f_2 = 1.1 GHz.

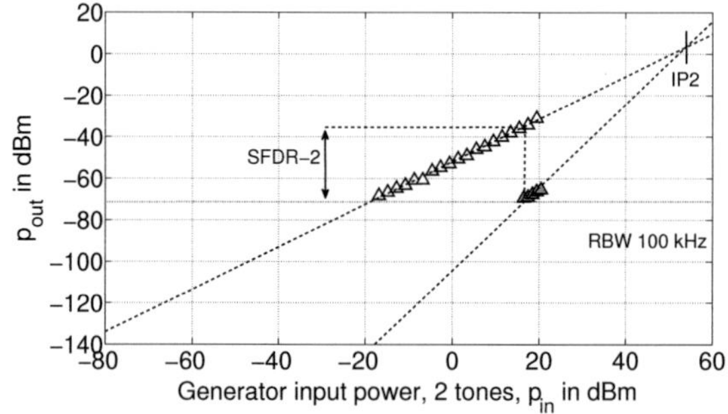

(b) Electrically amplified link, measured in the upper sideband. Two-tone signal: f_1 = 1 GHz, f_2 = 1.1 GHz.

Figure 3.15: Two-tone test for the direct modulation link

decrease of about 4.5 to 5 dB. A possible explanation is however the fact that during the measurement, the mediocre stability of the MLLD results in a (mechanical) repositioning that might cause deviations in signal power of up to optical 1 dB (or electrical 2 dB). For input powers > 15 dBm, second order modulation products appear; the equivalent IIP2 was found at 55 dBm. As we do not see

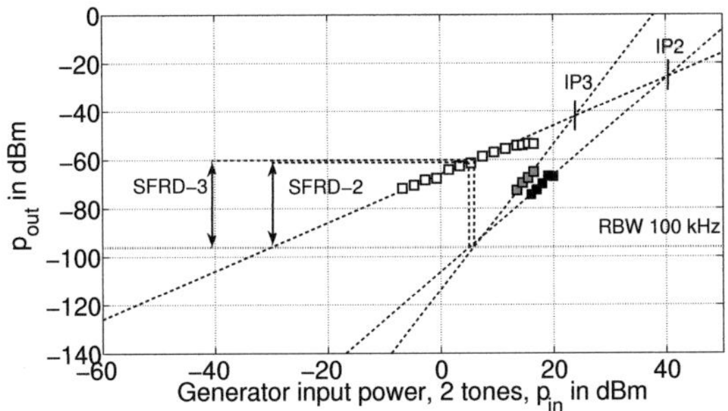

(a) Optically amplified, measured in the upper sideband.

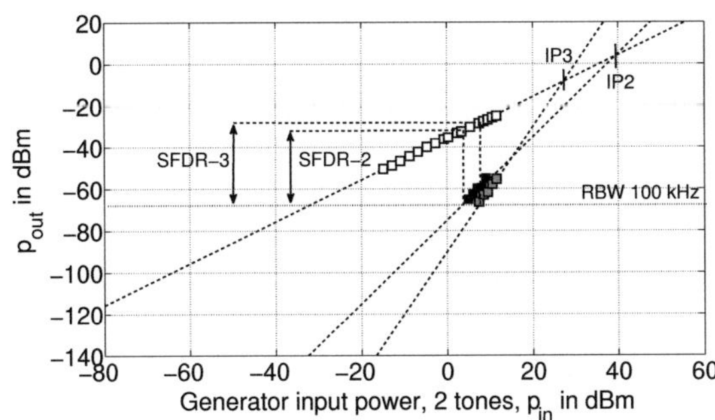

(b) Optically and electrically amplified link, measured in the upper sideband.

Figure 3.16: Two-tone test for the external modulation link

this effect in the intrinsic link, it is likely to be a distortion effect caused by the (relatively) powerful carrier signal which drives the RF amplifier into saturation although the sideband power is far from the compression point.

Optically amplified external modulation link: the external modulation link includes an EDFA and a VOA that have been set such that the resulting optical power incident on the photo-detector remains the same. The improvement we

Table 3.2: Summary of link performance: (*) dynamic range limited by max. input power for stable mode-locked operation, (X) cannot be measured.

Parameter	Direct modulation		External modulation	
	Intrinsic	Amplified (el.)	Amplified (opt.)	Amplified (el., opt.)
$P_{out,max}$ (dBm)	**-56.4**	**-32.3**	**-50.0**	**-26.0**
G_i (dB)	-65.0	-65.0	-58.6	-58.6
G_{total} (dB)	-71.4	-47.3	-65.0	-41.0
NF_i (dB)	65.0	65.0	58.7	58.6
NF_{total} (dB)	65.0	69.90	58.7	63.51
SNR_{max} (dB), RBW 100 kHz	**40.0**	**37.9**	**46.0**	**41.1**
IP1dB (dBm)	**x**	**x**	**10.4**	**10.9**
IIP2 (dBm)	x*	55.0	40.5	39.5
IIP3 (dBm)	x*	x	24.0	27.5
SFDR-2 (dB in 100 kHz)	x*	35.7	35.0	35.4
SFDR-3 (dB in 100 kHz)	x*	x	36.0	39.4
Noise Level (dBm/100kHz)	**-96.4**	**-70.2**	**-96.0**	**-67.1**

observe in output power (-50 dBm) is therefore only due to the increase in modulation efficiency when externally modulating the MLLD. The intrinsic link gain as measured was -58.6 dBm. The IP1dB for external modulation is found at 10.4 dBm. TOI now severely limits the dynamic range to 36 dB in 100 kHz, or 69.3 dB·Hz$^{2/3}$, with a noise level of -96 dBm/100 kHz, and an IIP3 of 24 dBm, which is considerably below the upper limit we have derived in section 3.2.3.2.

Optically and electrically amplified external modulation link: the inclusion of the electrical amplifier brings the output power up to reach -26.0 dBm. With a total link gain of -41 dB and an SNR ratio of 41.1 dB, this configuration seems preferable. The limiting intermodulation product is again TOI with an IIP3 of 27.5 dBm at a noise level of -67.1 dBm/100 kHz. The dynamic range resulting therefrom is 39.4 dB in 100 kHz, or 72.7 dB·Hz$^{2/3}$. It is not entirely clear why the inclusion of the electrical amplifier should result in an effective enhancement of dynamic range. A possible explanation could be the non-coherent summation of third order intermodulation products caused before and after a mixing stage, and before and after the electrical amplifier.

In conclusion, we find that the last configuration is possibly to be preferred, as it clearly allows for maximum output power at a reasonable SNR. On the other hand, it must be acknowledged that any external modulation link is always limited by the distortion inherent in its non-linear transfer function, and with respect to the requirements made for dynamic range in mm-wave systems in section 1.1.2.2, MLLD-based external modulation links still fall short of the target values of 77 - 80 dB· Hz$^{2/3}$.

3.2.4 Link frequency response

We will now give a justification for selecting the upper sideband from the two seemingly identical bands. In the direct modulation link as implemented, the following components determine the frequency response of the link: first, the intrinsic modulation bandwidth of the laser, second, the matching circuits of the MLLD and the photo-detector, and third the electrical components, i.e. the electrical amplifier and the down-conversion mixer. In figure 2.16(b), a typical modulation bandwidth measurement of the MLLD was shown for the unmatched MLLD. At

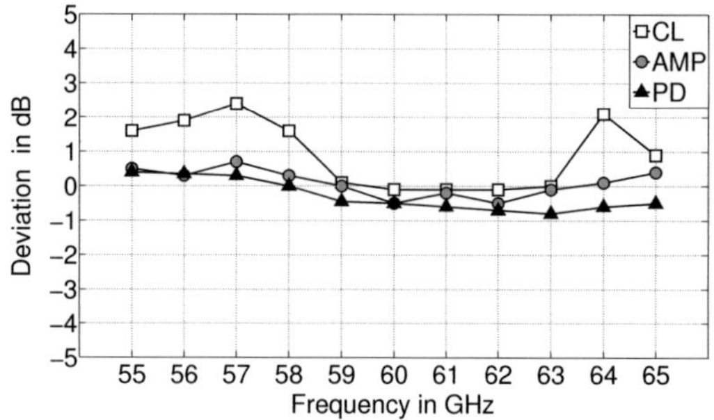

(a) Frequency-dependent deviation from specified value: gain of electrical amplifier (AMP, circles), photo-detector (PD, triangles) and conversion of down-conversion mixer (CL, squares)

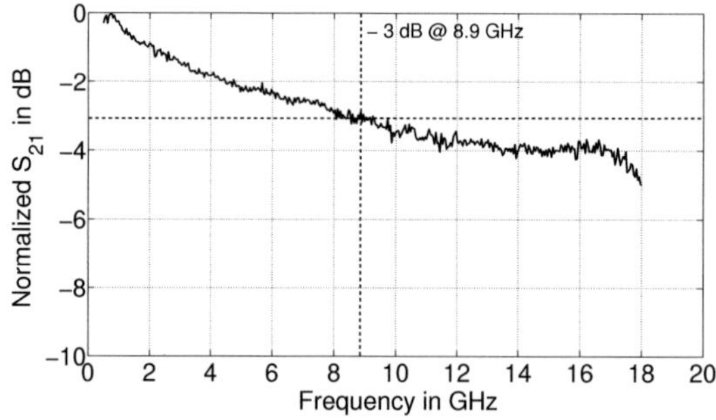

(b) Measured bandwidth of MZM

Figure 3.17: Frequency responses of link components

the output of the system, the offset frequencies are shifted by the beat frequency ν_F of the laser, i.e. 58.63 GHz in the case of L34. The cut-off frequency of 3.8 GHz is thus translated to 62.43 GHz for the upper sideband, or 54.83 GHz for the lower sideband. The electrical amplifier used for the experiment has a gain of 24 dB with a flatness of $\pm$ 0.5 dB across its operating range from 57.0 GHz to

66.0 GHz. Both down-conversion mixer and photo-detector have a flat response from 59.0 GHz to 63.0 GHz, with a frequency-dependent deviation from their specified datasheet values of < 0.5 dB. This deviation has been measured for all three components. All measurements were normalized to their specified values in the datasheet (mixer: conversion loss of 4.9 dB, PD response: -1.5 dB). If we limit our investigation to the upper sideband (flat region), we observe that the output bandwidth of the system is only limited by the MLLD response.

In the external modulation link as implemented, the frequency response is determined by the bandwidth of the MZM rather than by that of the MLLD, while all other components remain the same. As we show in figure 3.17(b), the -3 dB cut-off frequency of the MZM is approximately 8.9 GHz[31]. The external modulation link is implicitly more broadband. Should we choose to modulate the MLLD externally using this MZM, the output operating bandwidth of the link would be limited by the responses of the photo-detector, and the electrical components.

3.3 Conclusions on chapter 3

In this chapter, we have introduced the concept of sideband mixing and shown that the cross-beatings between modes and sidebands do add coherently to form a double sideband spectrum in the electrical domain. The impact of the low-pass filter effect due to the band-limitation of the photo-detector on the DSB spectrum was discussed. It is the principal cause for the high negative link gain in both direct and external modulation link and can only be mitigated by sharply limiting the number of modes to be modulated. This result confirms the findings of chapter 1 where we have seen the same effect on the carrier.

We have furthermore compared direct and external modulation of the MLLD. For the latter, mm-wave signal generation and E/O conversion functions are realized separately. It is therefore possible to circumvent the detrimental effect of the low-pass filter effect in an external modulation link by placing an optical filter and an optical amplifier before the modulator.

---

[31] The nominal bandwidth according to the supplier's datasheet is 18 GHz.

A comparison of the modulation efficiencies and the intrinsic link gain resulting therefrom has shown that external modulation is more efficient and has higher link gain, lower noise figure and provides a higher SNR than direct laser modulation, as long as a modulator with a small V_π value is chosen. In the external modulation link, we avoid the inherent trade-off between laser bias for maximum stability in terms of frequency jitter and phase noise, and laser bias for linear modulation. If it were possible to maintain stable mode-locking for lower currents where better slope efficiency is available, direct modulation efficiency could be increased so as to compete with external modulation.

For both modulation techniques, the frequency responses of the links are determined by the E/O converters. The electrode impedance of the commercial external modulator is source-matched over a wide range of frequencies, and the MLLD-based external modulation link is therefore inherently more broadband. A similar effort needs to be made for the (reactive) matching of the MLLD with regard to the power efficiency of the link. At the same time, we have seen evidence for the fact that the mode-locking operation in the cavity itself does not support the injection of very high modulation powers. Therefore, we cannot assume a considerable improvement of dynamic range by matching.
For intrinsic link gain $\ll 1$, the link noise figure is always limited by its reciprocal value. It naturally increases when electrical amplifiers are used. In the short-to-mid run, electrical amplifiers will however be necessary, as the MLLD-based link cannot yet provide the RF power levels needed at the BS. For a low-noise MLLD-based link, particular effort should therefore be devoted to increasing intrinsic link gain.

The suggestions for device design made in chapter 2 have been shown to be valid by the system-level analysis: a reduction of modes, as well as improving the chip-to-fiber coupling efficiency directly improves the slope efficiency, which in turn is the principal factor in determining link and, ultimately, link gain. Such modifications are left to the device designer. With the state-of-the-art MLLD, the system designer will probably opt for external modulation, even if this means an increased hardware effort. Throughout the chapter, we have assumed that the length of the direct or external modulation link was optimized for dispersion, but have not discussed what this implies. Chromatic dispersion, as the most crucial propagation effect in the MLLD-based fiber link, will be discussed in the following chapter.

4 Propagation Effects

Signal propagation in optical fibers is inevitably influenced by the physical properties of the fiber, which remains - despite its excellent transmission characteristics - a lossy and dispersive waveguide.

In this chapter, we will focus on the linear degradation mechanisms that occur in single-mode silica fiber when the MLLD signal propagates: attenuation due to absorption in the fiber material and scattering, and chromatic dispersion due to the difference in group velocities of different spectral components. For this reason, chromatic dispersion is also referred to as *group velocity dispersion* (GVD). After a theoretical introduction, we will examine through simulation the different phenomena related to mm-wave carrier generation in the MLLD-based system. In particular, we will look at the spectral width, the laser chirp, the shape of the optical spectrum and the phase noise on the optical spectrum, and we will compare those to experimental findings. We will see that for MLLD-based link design, we have to accept a dispersion-induced power penalty for carrier and modulation signals for certain link lengths. Transmission quality in terms of error-vector magnitude - a concept thoroughly discussed in chapter 6 -, as well as bandwidth-distance performance of the system, will be considered. It will be shown that GVD does not limit the absolute transmission distances in an MLLD-based RoF system, yet it needs to be carefully taken into account in link design.

Signal transmission can obviously only take place through the guided modes. Thereby, the notion of a *guided mode* refers to a spatial distribution of the optical field that can propagate through the fiber without changing its shape. Throughout this work, standard single-mode fiber is employed, which implies that only the fundamental spatial mode[1] HE_{11} can propagate. If both linear polarizations

---

[1] In fiber-optic communications, the fundamental mode is often denoted LP_{01}, for "linearly polarized" [Agr97].

are considered, the SMF supports two orthogonally polarized modes[2]. A great advantage of using SMF is that other forms of dispersion, such as modal dispersion, can be excluded because the energy is transported by a single mode only. Likewise, we can exclude from our discussion effects of polarization-mode dispersion as it is negligible compared to the dominant effects of GVD if we work at short transmission distances[3]. We will also disregard non-linear effects like Brioullin-scattering, Raman-scattering, four-wave mixing, self-phase modulation and cross-phase modulation, simply because the low power levels inside the fiber (< 10 mW) do not involve non-linear effects. A full discussion of these phenomena can e.g. be found in [Agr05].

In order to understand the effects of dispersion and attenuation in a MLLD-based RoF link, let us first consider a signal consisting of three modes[4] as illustrated in figure 4.1. At the photo-detector, two electrical beat signals will be generated, namely, the beating of the carrier with the lower sideband, and the beating of the carrier with the upper sideband. If both beat signals are in phase, they constructively interfere to build the electrical RF signal. If the signal is transmitted over a fiber link, each spectral component experiences a different phase shift depending on the link distance, on the modulation frequency and on the dispersive properties of the fiber. The phase shifts transfer to the beat signals. The formerly constructive interference becomes (partly or totally) destructive depending on the phase difference between the two tones, which causes a power degradation in the composite electrical signal [SNA97]. Across the link, periodic drops in RF power are thus to be expected.

For MLLD-based carrier generation, the same problem occurs *both* with respect to the mm-wave carrier *and* with respect to the modulation sidebands. This effect has been observed experimentally as early as in the mid-1990s [ANWL97], and

---

[2]It is important to note that with respect to the fiber, we speak of transversal modes, while the laser signal propagating across the fiber consists of numerous optical modes at wavelengths corresponding to the longitudinal modes of the laser cavity.

[3]For comparison, the dispersion coefficient associated to GVD is in the order of 17 ps/(nm·km), whereas the one for polarization-mode dispersion is about 0.1 ps/$\sqrt{\text{km}}$ [Cor12].

[4]The same situation occurs for a conventional ODSB system where the three spectral components are the optical carrier and two sidebands resulting, for example, from the intensity modulation of a laser diode.

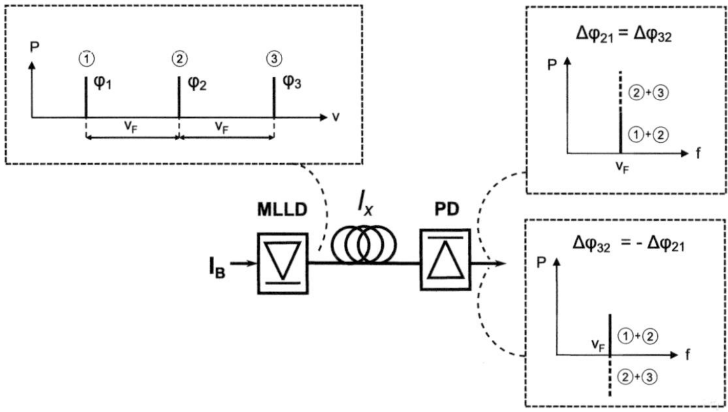

Figure 4.1: Dispersion effects on the photo-detected beat signal of three optical modes

was published again more recently with specific regard of passively mode-locked quantum-dash laser diodes [SBC$^+$10]. Yet a systematic study of dispersion-related aspects of mode-locked laser based carrier generation and its implications for mm-wave RoF links had not been published before the author's contribution [BPC$^+$11b], of which the following discussion is an extension.

4.1 Derivation of GVD in MLLD-based systems

The propagation of the center frequency component of an optical field in SMF at a carrier frequency v_0 is determined by its propagation constant β_0 [Agr05]. In the SMF, $\beta_0 = (2\pi \bar{n} c_0)/v_0$ can be determined with the help of the *effective refractive index $\bar{n}$* whose value lies in the range $n_1 > \bar{n} > n_2$, where n_1 and n_2 are the refractive indices of the core and the cladding of the fiber, respectively.

When another spectral component at a frequency $v = v_0 - \Delta v$, shifted from the center by Δv, propagates in the dispersive fiber, it assumes a propagation constant β_p different from β_0; it acquires an additional phase shift with respect to the center component. The different propagation velocities result from the frequency dependence of the effective refractive index. In general, its exact form is unknown. This is why β_p is usually written in its phenomenological form,

$$\beta_p(v) = \beta_L(v) + \beta_{NL}(v_0) + \frac{j\alpha(v_0)}{2}, \tag{4.1}$$

where $\beta_L(v) = \bar{n}(v)2\pi v/c$ is the linear part, $\beta_{NL}(v)$ is the non-linear part that we shall hence neglect, and α is the attenuation constant of the fiber. If the spectrum of the optical field is narrow compared to the carrier frequency - we have made the same assumption before[5]-, the frequency dependencies of the two latter factors can be neglected over the signal bandwidth. $\beta_L(v)$ can then be expanded in a Taylor series:

$$\beta_L(v) \approx \beta_0 + 2\pi\beta_1(v) + \frac{(2\pi)^2}{2}\beta_2(\Delta v)^2 + \frac{(2\pi)^3}{6}\beta_3(\Delta v)^3 \tag{4.2}$$

and

$$\beta_m = \left.\frac{d^m\beta}{dv^m}\right|_{v=v_0}. \tag{4.3}$$

The parameter β_1 is related to the group velocity of the pulse as $\beta_1 = 1/v_g$, while the parameters β_2 and β_3 describe the second and third order dispersion.
β_2 is known as GVD parameter (expressed in units of s^2/m); it is also related to the dispersion parameter D that includes both material dispersion and waveguide dispersion. It is commonly found quoted in fiber datasheets (expressed in units of ps/(km·nm)):

$$D = \frac{d}{d\lambda}\left(\frac{1}{v_g}\right) = -\frac{2\pi c}{\lambda^2}\beta_2. \tag{4.4}$$

The physical meaning of β_2 and D is the following: when the variables assume non-zero finite values, the medium is dispersive and spectral components propa-

---

[5]See section 2.1.1.1: the center frequency v_0 is of the order of about 194 THz at a wavelength of 1.55 μm, while the width of the spectrum Mv_F is of the order of 2 THz (i.e 30 - 40 modes separated by about 60 GHz). Thus, Mv_F is a small variation around v_0.

gate at different velocities. In standard SMF above zero dispersion wavelength (i.e. about 1300 nm), the dispersion parameter D is greater than zero ($\beta_2 < 0$), and the red components of a pulse (longer wavelengths) travel slower than the blue components (shorter wavelengths). The instantaneous frequency of the pulse envelope decreases with time. Therefore a pulse, i.e. the time-domain representation of a signal, broadens upon propagation on standard SMF. However, its spectral characteristics cannot change in a passive medium such as the fiber. The pulse must therefore be chirped. For $D > 0$ ($\beta_2 < 0$), the pulse is said to be *negatively* chirped ("red" chirp, down-chirp). For standard SMF and for operation in the C-band (1530 nm to 1565 nm), D is usually in the range of 16 - 19 ps/(km·nm) [GT00].

Third-order dispersion effects can be neglected in practice as long as β_2 is not close to zero (which is not the case for standard SMF), and as long as the duration of the optical pulse is not shorter than 5 ps [Agr05]. For the MLLD, the pulse duration can be estimated[6] as $\tau = 1/\nu_F$, which yields a value of about 16.7 ps for the MLLD employed here. Third-order dispersion is therefore not regarded in the following.

Other than dispersion, loss due to the attenuation of the fiber is another fundamental limiting factor. Its effects are not quite as serious for the relatively short RoF links, but attenuation will briefly be considered here. For an optical power $P_{\text{opt,in}}$ at the input of a fiber link with length l, the power recovered at the output is given by

$$P_{\text{opt,out}} = P_{\text{opt,in}} \cdot \exp\left(-\alpha l\right). \tag{4.5}$$

The optical power thus decreases exponentially with α describing the attenuation due to absorption in the material and scattering losses. α depends on the wavelength of the transmitted light. It is usually expressed in units of dB/km,

$$\alpha = -\frac{10}{l} \log_{10}\left(\frac{P_{\text{opt,out}}}{P_{\text{opt,in}}}\right). \tag{4.6}$$

---

[6]Strictly speaking, this holds only for unchirped pulses referred to as *Fourier transform-limited* pulses [GT00].

For SMF, a value of 0.2 dB/km was found as minimum attenuation [MTHM79] which is close to the fundamental limit for silica fibers (0.15 dB/km) [Agr97].

Optical fibers cannot generally be treated as linear-time invariant (LTI) systems. Under the assumption that the RoF system occupies a narrow bandwidth in the optical spectrum compared to the optical carrier frequency, the SMF can however be modeled as an LTI system with a transfer function H_{SMF} ([EWAD88], [GNN96]),

$$H_{\mathrm{SMF}}(v) = \exp\left(j\frac{\pi D l \lambda_0^2 v^2}{c}\right). \tag{4.7}$$

In the frequency domain, the optical field at the output of the fiber can be calculated as follows:

$$E_{\mathrm{opt,out}}(v) = \exp\left(-\frac{\alpha}{2}l\right) \cdot H_{\mathrm{SMF}}(v) \cdot E_{\mathrm{opt,in}}(v). \tag{4.8}$$

Applying H_{SMF} implies that one considers the propagation of different spectral components independently and linearly adds their powers to obtain the overall output power.

Dispersion effects can also be observed in the time domain: We have described the MLLD signal in equation (2.18); the following considerations are based on this description.

When the optical field propagates along the fiber, the phases φ_{m} of the comb will be modified across the fiber length l according to a wavelength-dependent propagation coefficient $\beta_{\mathrm{L}}(v)$:

$$\begin{aligned}
e_{\mathrm{opt,out}}(t) &= e_{\mathrm{opt,in}}(t) \cdot \exp\left(-j2\pi\beta_{\mathrm{L}}(v)\,l\right)\exp\left(-\frac{\alpha}{2}l\right) \\
&= \sum_{m=1}^{M} |A_{\mathrm{m}}| \exp\left(j2\pi v_0 t\right)\exp\left(j2\pi \cdot m v_{\mathrm{F}} t\right)\exp\left(j\varphi_{\mathrm{m}}\right) \\
&\quad \cdot \exp\left(-j2\pi\beta_{\mathrm{L}}(v)\,l\right)\exp\left(-\frac{\alpha}{2}l\right).
\end{aligned} \tag{4.9}$$

If we are only interested in determining the dispersion effects on the electrical beat spectrum, the term describing the optical carrier frequency $\exp(j2\pi v_0\, t)$ can be factored out without any loss of generality. If we also factor out all contributions of $\beta_L(v)$ which are not linked to chromatic dispersion, we obtain equation (4.10) for the remaining RF field after propagation through the dispersive fiber:

$$
\begin{aligned}
e_{\mathrm{out,RF}}(t) = {}& \exp\left(-\frac{\alpha}{2}\,l\right) \\
& \cdot \sum_{m=1}^{M} |A_m| \exp\left(j2\pi m\, v_F\, t\right) \exp\left(-j\frac{(2\pi)^2}{2}\beta_2(m\, v_F)^2\, l\right) \exp\left(j\varphi_m\right).
\end{aligned}
$$

$$(4.10)$$

The new mode phases φ'_m are thus:

$$
\varphi'_m = \varphi_m - \frac{(2\pi)^2}{2}\beta_2(m\, v_F)^2.
\tag{4.11}
$$

The electrical field is recovered by the (non-band-limited) photo-detector, yielding equation (4.12) where m and n represent at each time the numbers of two arbitrary interfering modes:

$$
\begin{aligned}
|e_{\mathrm{out,RF}}(t)|^2 = \exp(-\alpha l)\cdot\Bigg(& \sum_{m=1}^{M}|A_m|^2 + 2\sum_{m=2}^{M-1}\sum_{n=1}^{m-1}|A_m||A_n|\cos\bigg(2\pi(m-n)v_F t \\
& -\frac{(2\pi)^2}{2}\beta_2(m^2-n^2)v_F^2 l + (\varphi_m-\varphi_n)\bigg)\Bigg).
\end{aligned}
$$

$$(4.12)$$

In equation (4.12) we now observe the direct influence of GVD on the different spectral components in the beat spectrum: Even if the modes are locked (i.e. $\Delta\varphi_{mn} = \varphi_m - \varphi_n$), the effective phase of the beat note depends on the fiber length l.

In section 3.1.1 we presented the coherent superposition of the phase-locked spectrum when optical modes and optical sidebands mix. When the signal propagates across the fiber, this situation is qualitatively different: the different optical modes (i.e. the contributors to the sum in equation (4.12)) interfere constructively or (partially) destructively as l varies, and both the mm-wave carrier at v_F and its two sidebands experience periodic drops in power. In the following, we will first discuss dispersion effects on the mm-wave carrier and then the effects on the modulation sidebands in section 4.3.

4.2 Dispersion-induced carrier power penalty

Based on the expressions derived in the previous section, we can determine the impact of GVD on RF power. Four phenomena are of interest for the dispersion-related behavior of the MLLD signal: first, the number of modes in a MLLD varies slightly (reflected in the variable M). Second, the MLLD modes are locked, i.e. the phase differences between the modes are constant though not zero; this is a consequence of the chirp of the laser. Third, its optical spectrum is not perfectly flat but consists of modes with variable power (as reflected in the amplitudes $|A_m|$). Fourth, optical phase noise might disturb the laser mode-locking. The chirp phenomenon can be demonstrated for 3 modes while the others are better highlighted using a broader spectrum. All four phenomena can be studied through simulation and experiment. For the simulations, we consider the transmission of an optical field across a link of variable length l, the field exhibiting a spectrum with $v_F = 60$ GHz and a variable number of modes M. The periodicity with which power maxima and minima repeat themselves depends on v_F and amounts to $l_p = 2040$ meters for 60 GHz. The following simulation parameters were chosen: $\lambda = 1.55$ μm, $D = 17$ ps/(km·nm), and $\alpha = 0.2$ dB/km. The recovered RF power was evaluated at 60 GHz, thus holding the contribution of the beating of each pair of neighboring modes (band-limited case).

4.2.1 Influence of laser chirp

The limitation of the optical spectrum to only 3 modes results in the problem illustrated in figure 4.1. The interference of two optical modes results in a single beat note whose frequency $v_F = v_2 - v_1$ corresponds to the separation in wavelength of the two optical modes, and whose phase $\Delta\varphi_{21}$ is given by the difference between their phases. Adding a third, equally spaced optical mode will bring about the appearance of a second beat note at the same frequency $v_F = v_3 - v_2$ and with a phase $\Delta\varphi_{32}$, as well as a third one at the double frequency. The phases $\Delta\varphi_{21}$ and $\Delta\varphi_{32}$ will be modified by chromatic dispersion, giving rise to the periodic cancellation of the composite beat signal at v_F.

In section 2.2.3.4, we have introduced the phenomenon of laser chirping: at the output of the cavity, the MLLD pulse is positively chirped. Regarding the phase relationship between the modes, this implies that the phase differences $\Delta\varphi_{m,m-1}$ are constant but different from zero. Although the nature of the chirp in the quantum-dash MLLD is not yet fully understood, we can observe its effect on the fiber, as the positive chirp will at a certain length compensate the negative chirp accompanying the pulse broadening in the linear dispersive SMF. Unlike in the case of self-phase modulation, a non-linear phenomenon in the fiber that is sometimes exploited for chirp compensation[7], no additional spectral components are created. However, the phase relationship between the modes is altered by the fiber dispersion such that the phase difference between modes $\Delta\varphi_{m,m-1}$ are zeroed[8]. In the case of a chirped transmission, we expect to recover the first RF power maximum at a length l_{cc} instead of at zero meters. In figure 4.2, the dashed line represents the recovered RF power for $v_F = 60$ GHz, where all three modes are initially in phase. The solid line in the same figure shows the recovered RF power when the three modes carry an initial phase shift due to the chirp phenomenon ($\Delta\varphi_{21} = \Delta\varphi_{32} = \pi$). It can be observed that the signal power reaches its first maximum only after $l_{cc} = 1020$ meters of transmission. This initial phase shift is precisely the effect of the laser chirp which first counter-acts dispersion until being fully compensated for at the power maximum.

4.2.2 Spectral width

In the case of an optical spectrum of M modes, the beat signal of each two neighboring modes contributes to the recovered signal at v_F. The M-modes approach can be considered a superposition of the 3-modes case. The chirp phenomenon

---

[7] Referred to as *soliton* transmission.

[8] We actually cannot confirm by measurement that the phase differences decrease to zero; however, they are surely minimized at this specific length.

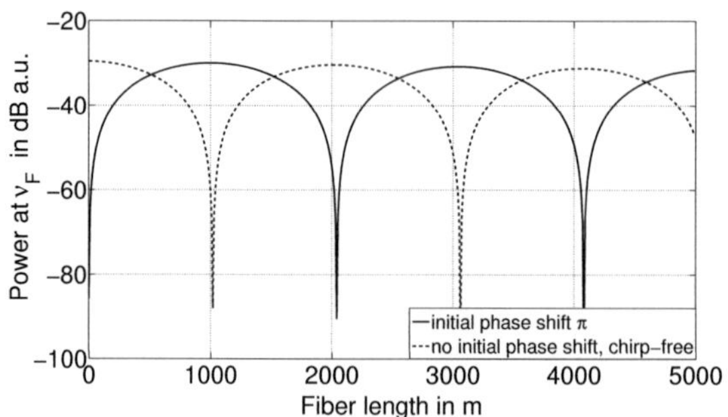

Figure 4.2: Influence of initial phase shift

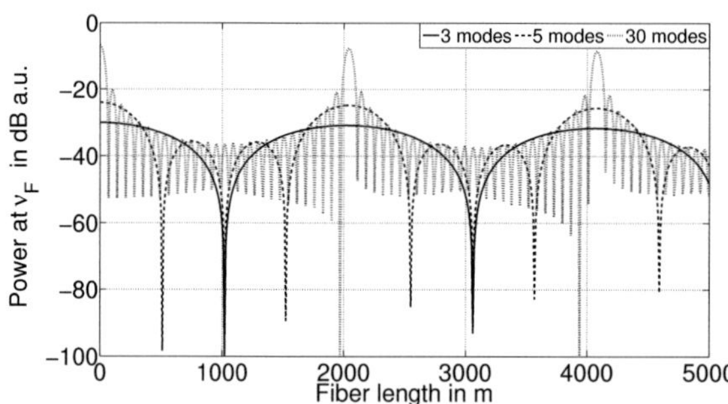

Figure 4.3: Appearance of local maxima for additional modes

shown in the previous section exerts the same influence as before on M optical modes. The subsequent addition of more modes of equal intensity results in the appearance of additional local maxima, while the previously observed principal maximum stays the same and the period l_p remains fixed at 2040 m, as shown in figure 4.3 for 5 and 30 modes respectively. All modes are assumed to have equal intensity and the phase relationship between the modes is constant $= 0$. However, the maximum lobe is narrowed as more modes are added. For the construction

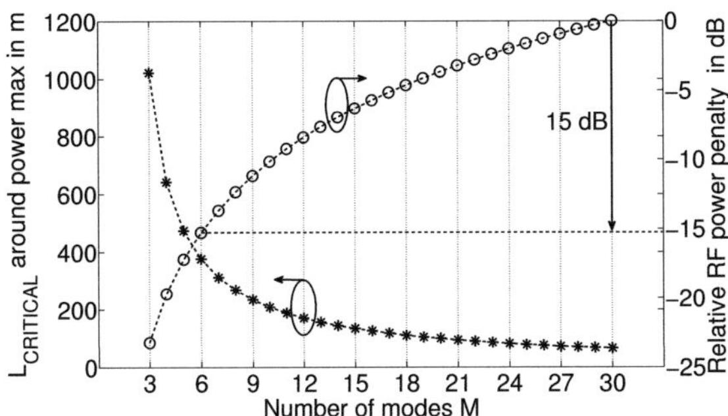

Figure 4.4: L_{CRITICAL} around power maximum at -3 dB

of high performance RoF networks, we would like to avoid the high sensitivity of RF power with link length. A figure of merit could be the *usable* fiber length L_{CRITICAL} around a power maximum, i.e. the fiber length margin around a power maximum if we allow a maximum power drop-off of ≥ 3 dB.

The relationship between the number of modes and L_{CRITICAL} is displayed in figure 4.4 (asterisks). From this relationship, we find one more reason to limit the spectrum to a minimum number of modes. If we introduced a filter in order to effectively limit the spectrum after the diode, a loss in optical power and in carrier power would be unavoidable. In figure 4.4, we have plotted also the carrier power penalty (circles) induced when limiting an initial flat spectrum of 30 modes to an arbitrary number of modes by means of a filter technique. If we want to limit the spectrum from 30 to 6 modes, we will lose about 15 dB in carrier power, but the usable fiber length along a power maximum is increased from about 70 to nearly 500 meters.

As discussed in section 3.2.1, carrier power is of limited interest as long as we can maintain a certain level in sideband power. We can avoid the power penalty on the sidebands if the filter is applied before the MLLD signal is modulated. For the current MLLD design, this is only possible when an external modulator is applied.

4.2.3 Shape of the optical spectrum

Another point to observe is the influence of the shape of the laser's optical spectrum. The MLLD exhibits a rather flat spectrum (3.1 dB / 10 nm), but even so, the gain curve and the repartition of cavity losses cause the spectrum to roll off at its edges. The outer modes do not significantly contribute to the beat signal power. It is therefore worthwhile to consider the influence of the shape of the optical spectrum on the recovered RF power at a given link length. In figure 4.5, two curves are represented for 30 modes each. While the solid line represents the RF power recovered from a field whose 30 modes exhibit equal intensities, the dashed curve represents the results from a measured spectrum whose modes vary with the laser's optical spectrum. These modes contain $\geq 95\%$ of the total power in the optical spectrum. When comparing the two curves in figure 4.5, it is observed that maxima and minima are considerably damped and thus reduced in power. Signal cancellation is consequently mitigated such that the local nulls disappear completely. Despite the power reduction in a maximum, this form of damping is not considered disadvantageous for the system as the usable fiber length L_{CRITICAL} tends to increase.

4.2.4 Phase noise on the optical modes

So far, the phase relationship between the optical modes has been considered perfectly constant. Yet, we have seen in section 2.1.1.3 that timing jitter, AM/FM conversion, and optical phase noise cause the modes of the optical spectrum and the beat notes in the electrical spectrum to assume Lorentzian shapes with a finite linewidth. In the context of GVD, the influence of random phase noise on the optical modes is therefore of interest. In figure 4.6 a curve for 30 modes, equal intensities and constant phase relationship (solid line) is compared with various other curves (dashed lines), the modes exhibiting an additional phase error. The phase error was modeled to follow a Gaussian distribution with a mean of $\mu = 0$ and a variance of σ^2.

Exemplary curves are shown for $\sigma^2 = 0.01$, $\sigma^2 = 0.04$, and $\sigma^2 = 0.16$. These values were chosen as they allow a clear demonstration of the effect of phase noise. It should however be noted that they are not related to actual measured noise pow-

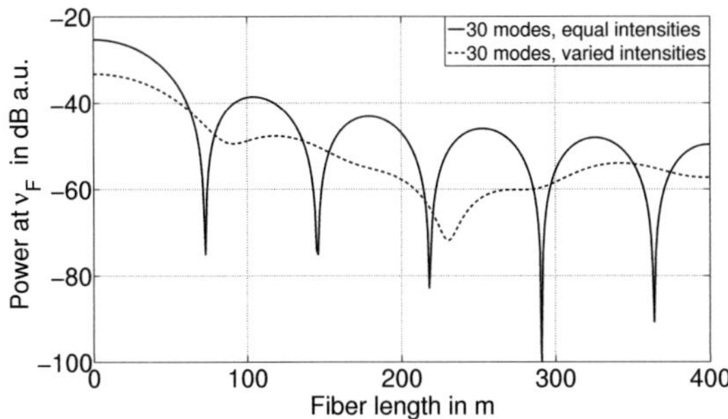

Figure 4.5: Influence of the shape of the optical spectrum, simulated for 30 modes

ers present in the MLLD. It is observed that the introduction of a random phase error already tends to dampen the effect of signal cancellation for small values of σ^2, while for higher values of σ^2, this effect becomes even more pronounced and the minima are shifted towards lower lengths. Note that the phase errors on different modes were assumed to be perfectly uncorrelated while for a real device, the discussion in section 2.1.1.3 suggests that the correlation between inner modes and outer modes typically varies, with the maximum correlation obtained at the center of the spectrum.

As will be presented in the following section, among the phenomena studied in this section, the chirp and the shape of the optical spectrum of the laser are found to be of particular importance for chromatic dispersion. While the effect of phase noise on chromatic dispersion might seem secondary with respect to the delivery of system power, it should be kept in mind that the phase noise of the optical modes remains a crucial parameter with respect to the linewidth of the recovered RF signal. It can thus nevertheless influence the system performance in terms of carrier stability.

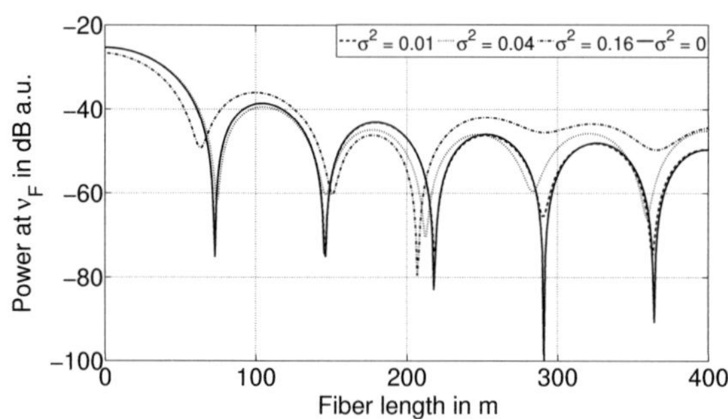

Figure 4.6: Influence of phase noise, simulated for 30 modes

4.2.5 Experimental validation

The measurement setup of the experimental validation corresponds to the one shown in figure 3.4 where the length of the SMF varies in steps of 12.5 m from 0 to 400 m. At each fiber connector, a loss value of 0.1 dB has been taken into account. All measurements were carried out on chip L34, and both direct and external modulation links have been tested.

In a first measurement, the relationship between link length and carrier power for different laser bias conditions is examined. The optical spectrum of the MLLD is shifted towards higher wavelengths at higher bias currents, while its flatness does not essentially change. The intensity of a certain mode is subject to vary slightly with the bias, determining the RF power that can be collected at a given length. Note that no data signal was transmitted and only the variation of the unmodulated carrier was studied.

Figure 4.7(a) displays the RF power recovered over the link length for a DC bias of $I_B = 200$ mA (black dots) and $I_B = 260$ mA (asterisks). These values were chosen as they result in stable mode-locking with a reduced RF linewidth for chip L34. The effect of the laser chirp is compensated for by chromatic dispersion at $l_{cc} = 72$ meters for both scenarios[9], indicating that the chirp does not depend

---

[9]Similar measurements on the second chip L872 have furthermore shown that l_{cc} is again in the order of 70 m.

strongly on I_B. It can further be observed that the recovered RF power follows roughly the same curve shape for short links, but varies significantly for link lengths greater than 250 meters. We attribute this result to the varied intensities of the optical spectrum, and have confirmed this assumption by comparing the simulated curves for both bias conditions to the measurements. In order to represent the measured scenario as closely as possible in the simulations, the optical spectra at both 200 mA and 260 mA were measured. The mode intensities were extracted and fed into the simulations. While the general behavior of the curves can be found using 30 modes, optimum coincidence with the measured values is found for a spectrum of 58 modes where the power ratio between the center and the edges of the spectrum is 30 dB. L_{CRITICAL} is found to be approximately 70 meters. Recalling figure 4.4, the value of 70 meters does indeed correspond to the value of L_{CRITICAL} of about 30 modes. As discussed in subsection 4.2.3, this circumstance can be attributed to the influence of the shape of the optical spectrum, indicating a strong contribution of 30 powerful central modes and a weak contribution of the outer modes due to the rolling-off of the optical spectrum at its edges. It is observed that the simulations coincide very well with the measured values, reproducing the principal maxima and the bumps that follow. In particular, the counter behavior of the two measurement series for a link length ≥ 250 meters is also recreated by the simulations.

While the laser chirp and the shape of the optical spectrum are indeed found to be relevant to the effects of chromatic dispersion, it is remarkable that the measured curves are closely reproduced without modeling phase noise. We thus draw the conclusion that an eventual phase jitter on the optical modes does not exert a marked influence compared to the variations of RF power induced by the variation of mode intensity. For that matter, the MLLD is thus sufficiently phase-locked. Figure 4.7(b) reveals a periodicity of $l_p = 2.11$ km (for chip L34: $v_F = 58.63\,\text{GHz}$). Optimum links therefore have lengths of $l_{\text{p,k}} = 72 + (k - 1) \cdot 2110$ meters, and $k = 1, 2..$ is the number of the maximum. The second maximum in our case is therefore found at $l_{\text{p,2}} = 2182\,\text{m}$.

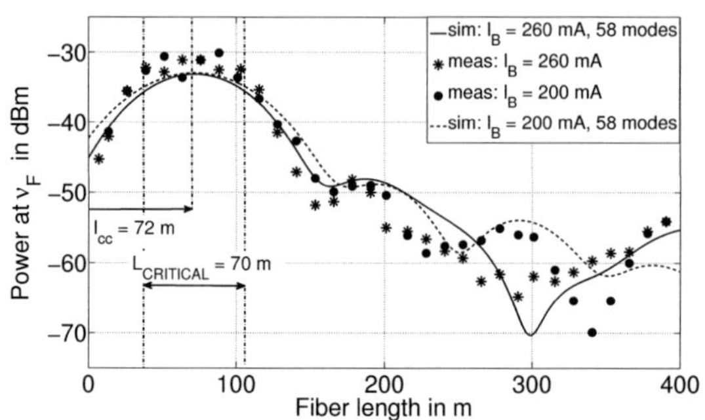

(a) 400m fiber length and varied bias conditions

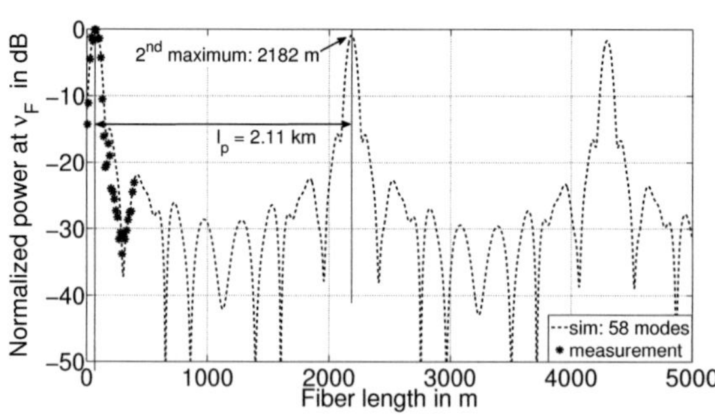

(b) 5km fiber length: Periodicity

Figure 4.7: Experimental validation of carrier power penalty, link based on chip L34

4.3 Bandwidth considerations

While the investigation on the carrier power penalty reveals a general fading behavior of the link, it does not yet allow conclusions to be drawn regarding the maximum available transmission bandwidth of the link. When the MLLD is modulated, the beat notes in the electrical spectrum corresponding to the inter-sideband and the mode-sideband beating will equally suffer a penalty induced by chromatic dispersion. The available transmission bandwidth of the link is of course determined by the penalty induced on the electrical sideband carrying the information.

The fading effect on the generated RF beat notes (i.e. beating between two modes, beating between a mode and a sideband, or beating between two sidebands) does not depend on the frequency generated by the beatings, but only on the wavelength distance between optical beatings in the *optical* spectrum that will superimpose at the same frequency in the *electrical* spectrum.

For a modulated spectrum, two different effects contribute to the overall impact of chromatic dispersion. Once again it is helpful to consider the case of a few modes, and subsequently generalize to M-mode operation.

First, if we are using an optical spectrum with only two modes, the resulting RF carrier signal (corresponding to the mode-mode beating labeled "1+2" in figure 4.8(a) linked to v_F) does not experience fading due to GVD. However, if this optical spectrum carries optical IM sidebands, the RF sidebands are again subject to dispersion-induced signal fading, and the fading periodicity is directly related to the wavelength distance between two beatings considered (corresponding to the intermediate frequency f_m): there are two mode-sideband-beatings labeled "1+2L" and "2+1U", separated by f_m, as shown in figure 4.8(a) which will generate the same electrical frequency at $v_F - f_m$.

For an intermediate frequency of $f_m = 3$ GHz, the periodicity induced is about 42 km (see figure 4.9) which shows a simulation of the variation of the slow-fading periodicity for center-frequencies from 500 MHz to 3 GHz. $f_m = 3$ GHz is the most severe case, for which we find a loss of less than 0.4 dB for distances < 1 km due to slow fading. The variation due to the change in intermediate frequency is < 0.01 dB for distances < 1 km. In figure 4.9, we can furthermore observe the RF loss induced by fiber attenuation. Assuming an attenuation coef-

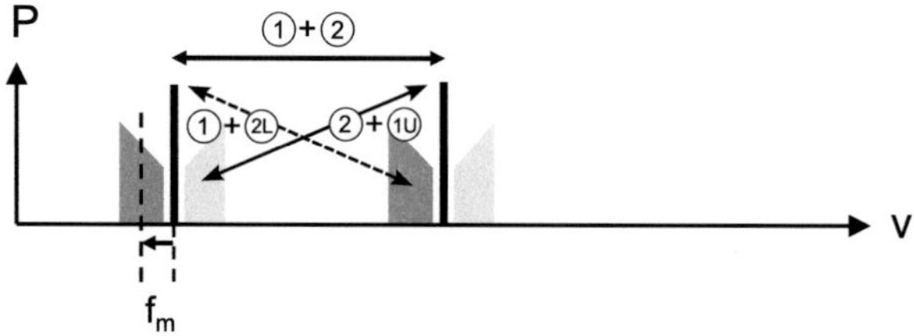

(a) Generation of the electrical lower sideband in the case of two optical modes, M = 2

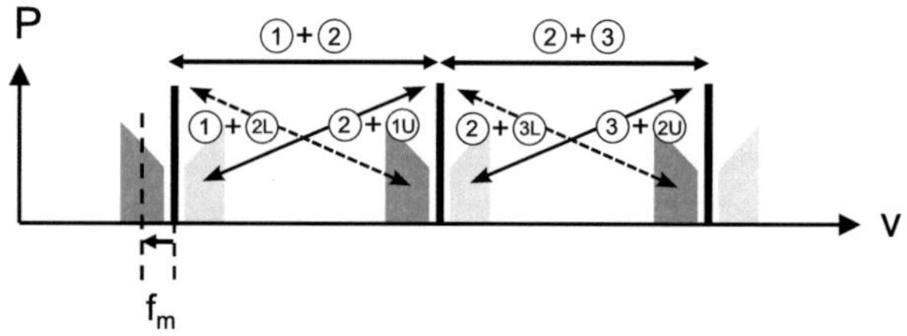

(b) Generation of the electrical lower sideband in the case of three or more optical modes, M = ≥ 3

Figure 4.8: Mode-mode and mode-sideband beating in the optical spectrum

ficient of $\alpha = 0.2$ dB/km in the optical domain, we expect an RF attenuation of twice its value; hence, at a distance of 100 km, the power loss due to attenuation is -40 dB.

Second, if a third (or further) mode is present in the optical spectrum, another type of fading appears that is directly related to v_F and caused by the mode-mode beatings labeled "1+2" and "2+3", see figure 4.8(b). As was shown in figure 4.7(b), the fast-fading periodicity is $l_p = 2.11$ km for $v_F = 58.63$ GHz. In this case, in addition to the previously considered chromatic dispersion effect, we will see a power drop-off every 2.11 km.

Other than these two effects, no other type of fading will occur. This can be explained by the following: The mode-sideband beatings labeled "1+2L", "2+1U", "2+3L" and "3+2U" in 4.8(b) will generate the lower RF sideband at an electrical

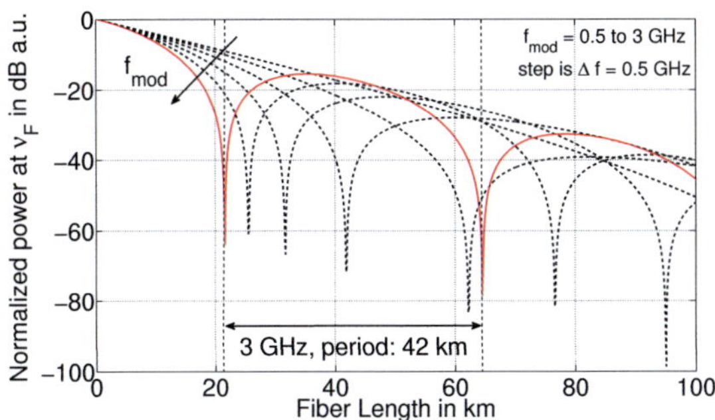

Figure 4.9: Slow-fading periodicity

frequency of $v_F - f_m$. However, the fading is not associated with this generated frequency but to the wavelength separation of the beatings in the optical spectrum. Let us first consider beatings "1+2L" and "2+3L". As these two beatings are separated by v_F (refer again to figure 4.8(b)) in the optical spectrum, the fading periodicity due to these beatings will be linked to v_F. The same is true for beatings "2+1U" and "3+2U" which are also separated by v_F. Thus, a complete and simultaneous fading can be observed for all 4 beatings for a fiber length associated with the mode separation v_F in the optical spectrum, independent from f_m. This description can be extended to a spectrum of M modes.

The two types of fading behavior will superimpose to give the overall fading behavior of the link: the (relatively) high v_F of 58.63 GHz will set a fast-fading periodicity of 2.11 km, modulated by a slow-fading periodicity of about 42 km for 3 GHz (or equivalent for other intermediate signal frequencies).

In conclusion, the link will principally be affected by the fast-fading behavior when we limit the transmission distance to a couple of hundred meters. It is therefore legitimate to examine the effects of GVD on a modulated spectrum in a *relatively narrow* bandwidth system. A test system available for such measurements is an IEEE 802.11a WLAN system with a channel bandwidth of 20 MHz. The following section will be dedicated to an experiment based thereupon.

4.4 EVM penalty in modulated links

In order to measure the GVD performance of the modulated link, the average error vector magnitude of the transmitted signal will be evaluated. EVM is a commonly quoted figure of merit for both transmitter and receiver evaluation. Reference values are defined in the communication standards (e.g., IEEE 802.11a [IEE99], ECMA 397 [Int08] and IEEE 302.15.3c [IEE09]), and the permitted values vary according to the data rate and modulation format implemented in the radio systems. While the received power in the modulation sideband, the noise level and the related SNR might be more intuitive criteria for the performance of a communication link, EVM allows for reliable diagnostics of other imperfections such as compression effects or phase noise. At the interface between the analog and the digital domains, EVM is measured in the constellation diagram after down-conversion to baseband. For the moment, we shall content ourselves with a definition according to which EVM is the root-mean-square (rms) value of the residual symbol error after an ideal version of the symbol has been subtracted. EVM will be discussed further in chapter 6.

As part of the dispersion analysis, we observe the degradation of EVM due to GVD in the fiber link. In the measurement setup depicted in figure 3.4, the spectrum analzyer is replaced by a heterodyne receiver system with a software interface capable of vector signal analysis[10]. Both direct and external modulation links were tested at the first maximum (expected around the chirp-induced offset l_{cc}), but not at the second one. The maximum distances that could be achieved for direct modulation were limited to about 130 meters, beyond which the power drop-off limits signal demodulation. Due to the higher received power and higher SNR for external modulation, measurements were possible up to 250 meters. Figures 4.10(a) und 4.10(b) show both the measured loss in power and the increase in EVM for direct and external modulation in a system employing chip L34. For a 24 Mbps 16QAM modulated signal, the received power in one sideband was measured (black circles), as well as the EVM after an averaging of 50 measurements taken subsequently (triangles). Note that the noise figure of the mixer was not available and has therefore not been taken into account for EVM calculation during vector signal analysis.

---

[10]We used Agilent's VXI-based VSA 89600 with the E2730/E2731 RF tuner module.

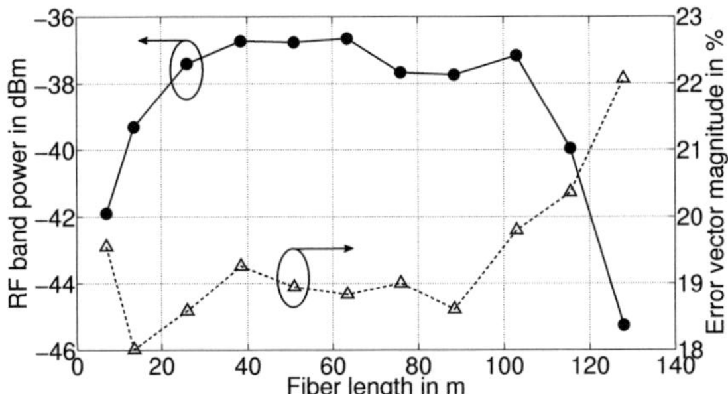

(a) Direct modulation. Test signal: IEEE 802.11a, 24 Mbps, 16QAM.

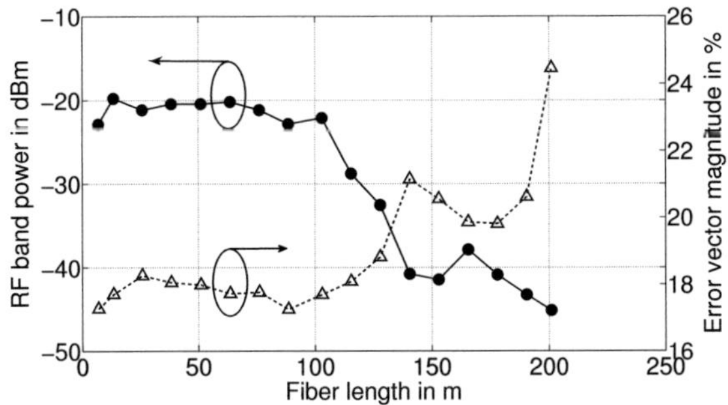

(b) External modulation. Test signal: IEEE 802.11a, 24 Mbps, 16QAM.

Figure 4.10: Measured received power and EVM performance of the modulated link, based on chip L34.

We observe that the external modulation link does not show the same power drop-off for low distances as the direct modulation link - or the carrier, for that matter. Again, this can be attributed to the chirp performance of the overall system. While intensity modulation of the MLLD typically further enhances the laser chirp in addition to what can be measured for unmodulated operation, the chirp induced by the MZM is typically low[11]. Other than that, the curves remodel the behavior that we have observed before for the carrier signal. The measurements suggest furthermore that the quality of transmission remains fairly constant over a width of about 80 to 100 meters which is reassuring for the connection of neighboring cells. As expected from the results of chapter 3, the external modulation link performs better also with respect to dispersion effects. The EVM floor that can be observed in both figures even for optimum link length ($\approx l_{cc}$) can be attributed to the frequency jitter and phase noise performance of the MLLD, an aspect to be discussed in the following chapters 5 and 6. In a low jitter (low phase noise) system, the EVM variation around the maximum might yet be more visible.

4.5 Conclusions on chapter 4

In this chapter, signal propagation in the MLLD-based RoF system was studied. As our investigation is limited to low power SMF transmission, the two relevant propagation effects are GVD and attenuation. While attenuation is not particulary problematical for the construction of short-haul data links, it nevertheless presents an RF power loss that cannot anymore be recovered.

Power drop-offs due to dispersion occur periodically; the *absolute* transmission distance that can be achieved in an MLLD-based RoF link is therefore not limited by GVD. As was shown in the previous sections, it is, however, limited to *specific* values of link length and a usable length L_{CRITICAL} around those values. Naturally, we are interested in maximizing L_{CRITICAL}. In the two previous chapters, we have presented reasons to limit the optical spectrum to a smaller number of modes. GVD is another reason, as L_{CRITICAL} increases with decreasing mode numbers. The findings of the simulations in figure 4.4 suggest that the number of MLLD modes should be limited to less than 30 in order to maintain a link length

---

[11] The absolute chirp value depends on the type of MZM (x-cut or z-cut crystals), as well as on the drive architecture (single electrode versus push-pull drive). For the single-electrode configuration used here, the chirp parameter is around 1 which is high compared to values of 0.1 - 0.5 for push-pull drive modulators. However, it is considerably lower than the chirp value for laser diodes (chirp of 2 - 3).

tolerance of at least 100 meters. This leaves a realistic margin for the construction of RoF networks. If a limitation of the spectrum cannot already be introduced at the stage of laser design, it is conceivable to filter a certain number of modes, e.g. by means of an optical filter that can then further reduce the number of modes. If the power loss is not compensated for by the use of an optical amplifier, the RF power level will decrease.

If optical filtering cannot be performed, the link length should be considered carefully when connecting a RoF cell to the system's CS. In the probable case of dealing with a moderate link length of several tens or hundreds of meters, the designer can quite simply use a fiber of l_{cc} (= 72 m) for all distances shorter than l_{cc}, thus work in the first maximum (shifted by the chirp-related offset length). For distances between l_{cc} and $l_{p,2}$, the designer must work in the second maximum using a fiber of length $l_{p,2}$. The fiber surplus can be stored in the CS. We have presented the relationship between the involved mm-wave frequencies, the fading periodicities and the tolerance to small length variations around a power maximum. Several parameters of the MLLD (optical spectrum, chirp, and the spectral width) play a critical role for dispersion performance.
We have furthermore shown that the fading effects present during the transmission of a modulated signal do not essentially limit the bandwidth-distance performance.

In a final remark, we address the problem of OSSB transmission. In single-wavelength ODSB systems, the problem of GVD or chromatic dispersion has been encountered with an OSSB technique which consists in eliminating one of two optical sidebands before transmission. In a narrowband system, this can be achieved by using a dual-electrode MZM in a so-called *push-pull*configuration, where the modulation signal is sent to an electrode with a phase delay of $\pi/2$ with respect to the signal feeding the other electrode [SNA97]. OSSB has also been shown in a 60 GHz RoF system using RHD and two cascaded MZMs. Here, a fiber Bragg grating (FBG) was used in order to filter the spectrum in such a way that an OSSB spectrum was obtained. This solution is preferable over the dephased push-pull configuration in the sense that it is easily applicable to broadband systems [NCP$^+$09]. These techniques are only applicable in the presence of one ([SNA97]) or two ([NCP$^+$09]) carriers.

In the case of the MLLD-based system, the RF signals are constructed from various superimposed beat signals. When the second sideband of a particular optical mode is eliminated, it will nevertheless appear in the RF spectrum through the beating of the remaining optical sideband with a higher neighboring mode.

5 Stabilization Techniques

In chapter 2, measurements performed on the MLLD-generated carrier signal have revealed a significant instability of the MLLD beat frequency ν_F. At offsets sufficiently far from the carrier, the resulting phase noise is measurable; it can be quantified by a value of about -55 dBc/Hz at 10 kHz offset. Through the mixing process inherent to the MLLD, the frequency jitter translates onto the modulation sideband and hence spoils the signal quality unless frequency tracking is implemented in the receiver. Even with frequency tracking, the phase noise inevitably corrupts the signal.

The problem of high phase noise in systems based on passively mode-locked laser devices has been explicitly addressed in the literature. It has been claimed before that due to the tight requirement on oscillator linewidth (several Hz to several kHz), a free-running passively mode-locked laser cannot successfully serve as optoelectronic oscillator [AMP00], although a record RF linewidth of 500 Hz has been shown for a quantum-dot MLLD pulsing at 10 GHz [CTPW09]. For mm-wave generators, however, measured linewidths are of the order of several 10 kHz to several 100 kHz. It is generally recognized that, unless it is possible to construct a stabilization architecture compensating for the phase noise of the free-running device, the MLLD cannot compete with its electrical counter-part - the voltage-controlled oscillator (VCO). The VCO exhibits mediocre phase noise characteristics when free-running, but it can usually be stabilized by a crystal-based phase-locked loop (PLL) [dB32]. Other than for the reduction of phase noise, the PLL is typically employed in synthesizer circuits, where the VCO can be tuned to a precise frequency. This property comes in handy when the communication system needs to be tuned to a specified radio channel with a given band center frequency, a feature the MLLD-based system *a priori* cannot provide, even if the MLLD beat frequency ν_F could be selected with high precision. It is thus highly desirable to find a method to both stabilize and tune the trans-

mission signal to custom frequencies. In a conventional PLL, the MLLD would be embedded in an optoelectronic loop. A fraction of the MLLD signal would be detected and routed back to the diode to close the loop. We have however already mentioned in chapter 2 that the construction of a conventional PLL is not possible in the case of the quantum-dash MLLD because we do not have access to a physical quantity effectively controlling the oscillation.
There are a number of other techniques that have been developed to stabilize the oscillation of a MLLD. A brief overview of those techniques will be given in the following section in which we will also describe why they are only of limited use in our context.

The main part of the chapter will then be dedicated to the demonstration of a novel architecture based on the PLL principle which allows frequency stabilization and tuning, as well as a reduction of phase noise in the MLLD-based link. We base our argument on the fact that the instability present in the carrier signal is of little concern as long as we can prevent it from translating onto the modulation sideband which represents the information to be transmitted by the system. The proposed concept is validated experimentally by means of a system demonstrator which was developed for this purpose.

5.1 Optical and hybrid stabilization techniques

Active mode-locking. The most common technique to stabilize a mode-locked laser is that of active mode-locking (see also section 2.1.2), where a high-stability electrical driving signal at the fundamental beat note of the laser or a subharmonic is applied directly to the laser. A phase noise performance of -70 dBc/Hz at an offset frequency of 5 Hz was shown for a two-section actively mode-locked laser where the driving signal was applied to the saturable absorber (SA) section[1]

---

[1]In this case, the bias voltage of the SA is modulated. The importance of applying a voltage signal to the SA section rather than a current signal to the gain section is explained in [AMP00]: (a) current modulation affects the laser gain only indirectly through the modulation of the carrier population. This process becomes inefficient at frequencies beyond the cut-off frequency of the intrinsic laser bandwidth. Voltage modulation of the SA instantaneously modulates the SA coefficient; and (b) the small length of the SA facilitates phase alignment.

[AMP00]. Active mode-locking allows for the stabilization of the laser at its self-pulsation frequency. It does not enable frequency synthesis. In an IM-DD system where active mode-locking is employed, the IF oscillator must then be tunable. This is not a problem *per se*, as a synthesizer loop can easily be used in the IF range. On the other hand, a stabilized electrical mm-wave oscillator is needed for active mode-locking, which is awkward when the MLLD is supposed to replace the electrical mm-wave generator[2].

Optoelectronic PLL. The optoelectronic PLL is a straightforward extension of the electrical PLL. It basically corresponds to a type of active mode-locking where the driving signal - fundamental beat note or subharmonic - is generated from the laser signal itself. A phase or frequency error with respect to a reference is detected, fed back to the laser and minimized by the loop through a fine adjustment of the laser beat frequency by active locking. Its application to MLLD was first reported by Buckman *et al.* [BGP$^+$93]. It typically includes an optical coupler and a photo-detector, so that a part of the laser light can be O/E converted and fed back to the laser. PLL techniques avoid the necessity of applying the mm-wave signal directly to the laser as in conventional active mode-locking [KT96]. However, they still require a very stable electrical signal at the desired fundamental beat frequency, or at a sub-harmonic. All optical PLL techniques featuring a direct feedback path to the laser must assume that the laser oscillation can be fine-tuned by a control signal. The analogous property in electronics is the approximately linear[3] relationship between tuning voltage and oscillation frequency in VCOs. As was shown in section 2.2.3, the relationship between the beat note frequency and the bias current - the only electrical control signal at our disposal - is not linear. On the contrary, mode-locking sets in for dedicated bias currents only, and it depends to a large extent on the mechanical fine-tuning in the air-coupled arrangement. Thus, stabilization by an optoelectronic PLL as proposed by [BGP$^+$93] and [KT96] is not feasible.

---

[2]Active mode-locking at a subharmonic did not yield the desired result as the quantum-dash MLLD in this work does not dispose of an SA section.

[3]This is a constraint we need to make if we want to model the system as an LTI system. For a servo loop to function properly, it is usually sufficient that the relationship between the two quantities be monotonically increasing. In a real PLL, the relationship between tuning voltage and oscillation frequency is typically not perfectly linear either. For LTI analysis it is, however, sufficient to linearize its transfer function in the entire frequency range of operation.

Optical injection. Injecting an optical signal at the fundamental beat note or subharmonic has also been shown to stabilize the MLLD output signal. The best results were obtained when the two injected modes were within a range of 0.001 - 0.01 nm of the two strongest modes of the MLLD [ALN+96]. The optical injection locking technique is considered highly sensitive and too complex for system-level integration in mm-wave RoF systems. It might however yield interesting results for the stabilization of a THz beat note, where even subharmonic electrical locking can be difficult.

External optical feedback. Stabilization can be achieved when a fraction of the laser light is coupled into an optical fiber and fed-back into the cavity at a well-defined level, using polarization controllers. Linewidths of 1 kHz or less have been obtained for quantum-dot MLLDs operating between 5 and 10 GHz, and for feedback levels typically ranging between -20 and -15 dB [MRA+09], [LGL+11]. For the mm-wave quantum-dash MLLD, these results could not be reproduced.

Coupled optoelectronic oscillator. Van Dijk *et al.* have integrated a quantum-dash MLLD (39.9 GHz) of the same type as the one used in this work in a coupled optoelectronic oscillator [vDEB+08]. In this setup, the laser light is routed through an optical coupler, split, and routed through a fiber interferometer, the two arms of which are of different lengths. The signals are separately detected, combined electrically, amplified and re-routed to the laser in a direct modulation configuration. The so generated electronic feedback has allowed a phase noise reduction of more than 15 dB in the low-frequency range, i.e. from -55 dBc/Hz to -70 dBc/Hz at an offset frequency of 10 kHz. Unfortunately, the setup is rather complex and sensitive to optical feedback into the laser, i.e. to retro-reflections from the fiber into the cavity.

In conclusion, conventional stabilization techniques have turned out to be either impractical or unrealizable for the quantum-dash MLLD. Except for the optical PLL, none of the proposed techniques can be used for frequency tuning. In the following, we propose a solution based on the insight that, in a double sideband (DSB) spectrum, the amount of carrier instability is of minor importance, *as long as the sidebands carrying the information can be stabilized.*

5.2 PLL-based modulation sideband stabilization

Of the above techniques, only the PLL-based methods allow both the for stabilization and the tuning of the oscillator signal. We will thus concentrate on finding a control loop architecture which is applicable to the MLLD-based RoF system, even though we have seen that conventional methods fall short of providing a solution. Some preliminary considerations are helpful. In the transmitter of a conventional AM radio communication system, the complex digital baseband signal (I, Q) modulates an analog IF signal at an intermediate frequency f_{IF}. This IF signal is subsequently up-converted to a dedicated frequency band by mixing it with a carrier signal at f_{LO}[4]. The upper and lower sidebands correspond to the mixing products of the carrier signal and the information signal at f_{IF}. They are centered around $f_{LO} + f_{IF}$. From the DSB spectrum, one sideband will be selected by means of a filter and transmitted by the antenna[5]. The error-free operation of the radio system can only be assured if the users can synchronize their receiver equipment with the central frequency of the transmitted sideband. The stability of this sideband is therefore essential for the overall functioning of the system. Noise in the carrier signal will translate onto the sidebands through the mixing process. In conventional radio systems, measures are usually undertaken to *stabilize the carrier signal*; through the mixing process, the sidebands are stabilized. Typically, the output frequency of the IF and LO sources are controlled and stabilized by means of a phase-locked control loop before up-conversion.

In the MLLD-based mm-wave RoF system, the complex baseband signal is first up-converted to f_{IF} range. Subsequently, it is E/O converted by an appropriate modulator or by the MLLD in a direct modulation configuration. Up-conversion to the mm-wave range is also performed by the MLLD. All phase noise induced by the MLLD will inevitably appear in the electrical spectrum. We recall that the electrical equivalent for the laser-based up-conversion is a mixing process, where

---

[4]In a direct-conversion transmitter, the intermediate frequency stage can be omitted; the carrier directly feeds the I/Q modulator.

[5]The method is inherently inefficient as half of the signal power is lost; this is one reason why DSB-AM is not usually employed anymore for electrical radio systems. In the RoF context, DSB-IM is however the state-of-the-art technique.

the desired sideband again corresponds to a mixing product of the fundamental beat signal of the MLLD at ν_F, and the information signal at f_{IF}. Instead of stabilizing the carrier signal - we have seen that this is not feasible using the MLLD - we can *let the control loop act on the mixing product* itself. We do so by controlling the output frequency of the IF oscillator at f_{IF}. The structure of a PLL operating on a mixing product of the VCO and a second oscillator is one relying on the principle of in-loop heterodyning; it is also known as "translation loop" from the literature [Ega99] where it has been applied to purely electrical systems. We will show that it is of great use also in the context of RoF systems. The translation loop will be discussed in the following section.

5.2.1 In-loop heterodyning: The up-conversion translation loop

A *down-conversion* translation loop (TL) is typically used in order to provide very high VCO output frequencies (e.g. in the mm-wave range) while at the same time reducing the operation frequency, i.e. the frequency of a reference f_{ref}, at the input of the phase detector. As shown in figure 5.1, the operation frequency of the phase-frequency detector corresponds to the difference frequency of the VCO ("VCO-LO") and the translation oscillator ("TL-LO") when their two signals are fed into a down-converting mixer. In the down-conversion TL, the operation frequency can be kept very small which is usually advantageous to the construction of the PLL because the divider ratio N can be kept very small. This is important as the divider ratio exhibits a strong influence on in-loop phase noise [Ega99]. A fundamental condition of this structure is that the translation oscillator be a particularly stable - i.e. low phase noise - oscillator, or that it be stabilized by a second PLL. If this is not the case - which is what we will assume for the following - any additional phase noise on the translation oscillator will enter the system through the mixing process and will inevitably appear at the output of the system. An analogous situation would occur if the PLL acted on the translation oscillator instead, and the then free-running VCO-LO would be further corrupted by noise.

In the presence of translation oscillator phase noise, the signal tapped at the output of the VCO-LO is unusable because the PLL acts on a source (VCO-LO)

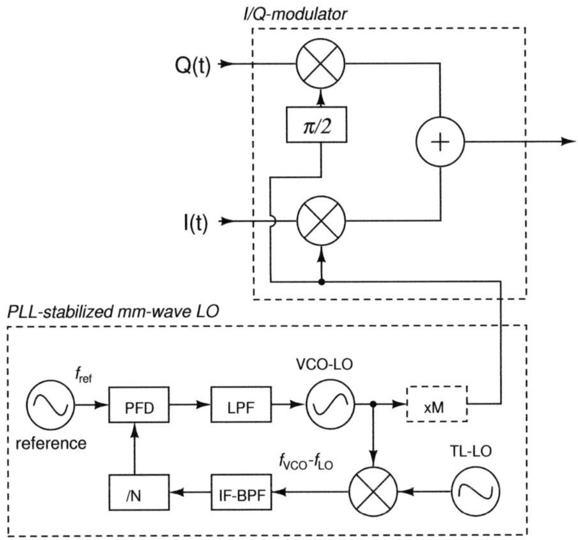

Figure 5.1: Conventional use of a translation PLL. The operation frequency of the PLL is shifted by heterodyning the VCO with a translating LO. A multiplying stage might be necessary for mm-wave operation.

which is not correlated to the true origin of the phase noise. For these reasons, the TL as known from the literature is not suitable for signal stabilization in the MLLD-based RoF system where the MLLD-PD pair assumes the role of the TL-LO. The underlying structure of the TL, however, serves as the basis for the novel stabilization architecture developed in this work. The proposed architecture is based on the following ideas which should be kept in mind when studying the following figure 5.2:

- The complete stabilization structure incorporates a PLL, as well as a photodetection stage and a carrier-recovery stage *on the transmitter's side.*

- The ensemble operates as an up-conversion translation loop where the translation oscillator signal is provided by the optical sub-system - in particular, the MLLD-PD pair. The PLL acts on the mixing product of the carrier signal provided by the MLLD and the signal provided by the IF oscillator.

- In the DSB-IM MLLD-based RoF system, this mixing product, i.e. the upper or lower sideband in the electrical spectrum, is the *desired* output signal of the system. The mixing product carries the linear[6] contributions from both the IF oscillator and the MLLD; regarding the phase noise present in the mixing signal, it is correlated to both sources. The PLL can thus be constructed either to act on the MLLD - which is *de facto* not possible - or on the IF oscillator.

- The mixing product in the DSB-IM spectrum exhibits a bandwidth. However, inside the loop, the phases of a single-tone mixing product and a stable reference must be compared. It is thus necessary to suppress the frequency bands around the center frequency $f_{RF} = f_{LO} \pm f_{VCO-IF}$.

Figure 5.2 depicts the system architecture. The optical sub-system is shown in the dotted-line box. While the optical signal is sent out to the RoF system on one optical branch, a fraction of it is immediately detected and routed to the PLL. The dashed-line boxes are schematic representations of the electrical spectrum which can be measured at the different points in the system. The so modified PLL provides a control signal for the VCO working in the IF range ("VCO-IF"). The VCO-IF outputs a signal at f_{VCO-IF} and feeds the I/Q modulator (a). Through I/Q modulation, the signal obtains a bandwidth (b). The IF signal is E/O converted; it modulates an optically generated mm-wave carrier signal, so that an IM-DSB spectrum is obtained at the output of the O/E converter (c). One of the sidebands will be selected for transmission. The selected sideband corresponds to the mixing product to be stabilized by the PLL centered at $f_{RF} = f_{LO} \pm f_{VCO-IF}$. However, it generally exhibits too large a bandwidth in order to be fed into the PLL directly. The frequency bands are removed in three steps. First, the carrier is recovered from the IM-DSB spectrum (d). This can be achieved by means of a narrow band-pass filter. Second, the IM-DSB spectrum is recreated by mixing

---

[6]in a first-order approximation

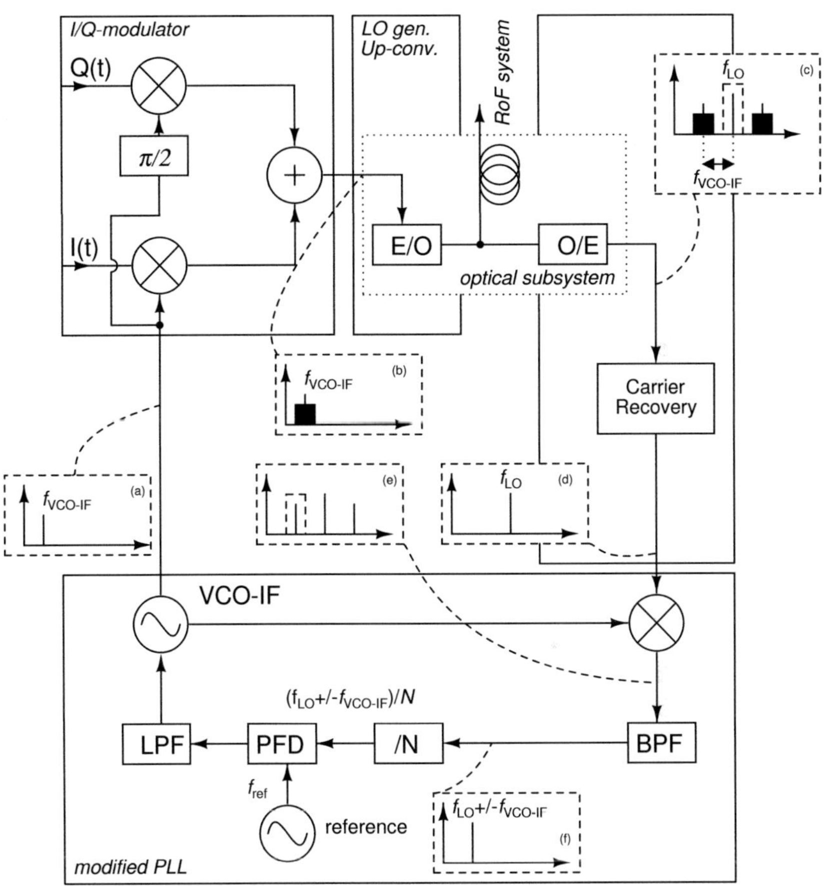

Figure 5.2: Novel sideband stabilization technique for RoF system: Upconversion translation loop in the RoF transmitter

the recovered carrier and the unmodulated VCO-IF signal (e). Third, the single-tone sideband is selected by another narrow band-pass filter at the output of the mixer (f). In principle, either of the two sidebands can be selected as long as the choice is taken account of in the design of the loop. The signal is then fed into a divider with ratio N, and a phase-frequency detector (PFD) where its phase is compared to that of a stable reference. The control signal is low-pass filtered in the loop filter (LPF, low-pass filter) which provides a DC control signal to the VCO-IF.

In the presence of phase noise on either the VCO-IF or the optically generated carrier, the total phase error of the mixing product is compared to a stable reference. Through the control of the VCO-IF, the PLL compensates for the total phase deviation. In principle, this architecture can be applied to any type of optical carrier generation. In particular, it can be applied to the directly or externally modulated free-running MLLD where the frequency jitter and the phase noise induced by the MLLD will be compensated for by the loop. This architecture and stabilization principle have been submitted by the author and her co-workers to the European Patent Office for patent protection [ZBCP].

5.2.2 The phase-locked loop: Methods of analysis and principle of operation

The continuous-time phase-locked loop is usually described using Laplace analysis[7] and the techniques of control theory. The PLL corresponds to a plant-feedback model acting on a state variable, namely phase or frequency. The PLL is considered an LTI system which can as such be represented by a transfer function $H(s)$. In the Laplace ("s") domain, the response $Y(s)$ of the system to an input $X(s)$ in the Laplace domain is then the product $H(s)X(s)$. The basic plant-feedback model is shown in figure 5.3(a), where the blocks $G_F(s)$ and $G_R(s)$

---

[7]The description of the PLL as an LTI system suffers notably from two drawbacks: first, the steady-state Laplace analysis is not able to predict the typically non-linear acquisition process of the PLL and can therefore only accurately describe the locked PLL. Second, a number of processes in a real PLL are discrete processes of time, and the sampling process can in general not be treated using Laplace analysis [HBMM04]. Nevertheless, Laplace analysis has proved an appropriate tool which yields sufficient accuracy for most applications. Therefore it will be the method of choice also in this work.

correspond to the (forward) open-loop transfer function, and the feedback (or reverse) transfer function, respectively. E is the error of a controlled variable C with respect to a reference R. If we consider the phase of the reference $\varphi_R(t)$ as the input variable, and $\varphi_C(t)$ as the output variable, the closed-loop transfer function of the loop $H(s)$ is

$$H(s) = \frac{\Phi_C(s)}{\Phi_R(s)} = \frac{G_F(s)}{1 + G_F(s)G_R(s)}. \tag{5.1}$$

where $\Phi_C(s)$ and $\Phi_R(s)$ are the Laplace transforms of the phase signals $\varphi_C(t)$ and $\varphi_R(t)$ respectively[8].

A conventional PLL as shown in figure 5.3(b) consists of three components in the forward path; their respective transfer function is given in the block diagram: a phase-frequency detector (PFD), described by its linear gain K_P, a loop filter with a transfer function $Z_{LPF}(s)$ (e.g. the two-terminal impedance for a passive loop filter implemented as low-pass filter), and the controlled oscillator with a tuning sensitivity (usually in MHz/V) described by a gain K_V. The s^{-1} block represents an integration, namely the relationship between the phase and the frequency domain in Laplace notation. In the feedback path, a divider with a ratio N is usually employed to match the frequencies of the reference and the controlled oscillator. The open-loop transfer function and the reverse transfer function of the conventional PLL are then:

$$G_F(s) = K_P \cdot Z_{LPF}(s) \cdot \frac{K_v}{s}, \tag{5.2a}$$

$$G_R(s) = \frac{1}{N}. \tag{5.2b}$$

A thorough discussion of classical PLL theory can be found in [Ega00], [Gar79] or [Ban06].

When describing the up-conversion TL, the open-loop und reverse transfer functions will need to be completed by the contributions of additional components, i.e. the **mixer** and the **band-pass filter**. As will be shown later, a third compo-

---

[8]For simplicity of notation, the explicit time-dependence of the phases φ is dropped in the following.

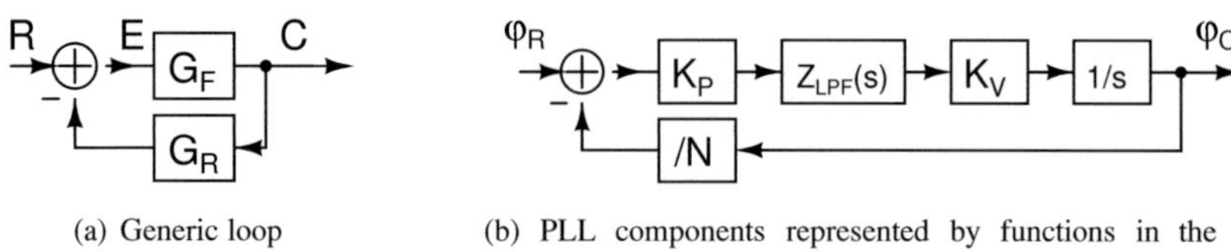

(a) Generic loop

(b) PLL components represented by functions in the s-domain

Figure 5.3: Classical PLL representations, from [Ega00]

nent, namely, an additional **RF amplifier,** will have to be included. Constraints on the linearity of the phase response of additional components can be relaxed for continuous-wave operation, as long as the phase shift introduced by a component is well-defined and can be taken into account in the design of the loop. Each component adds a contribution to the overall transfer function, but also to the system noise figure. In the following, the general implications of introducing any of the three components will be revealed.

5.2.2.1 Effect of the mixer

At first sight, the mixer seems to pose a problem as it is evidently a non-linear component implying a multiplication of two signals, while the Laplace transform is a linear transformation. Regarding the phase or the frequency of a signal, the mixing process is however a linear translation that can be integrated into a closed-loop transfer function for phase. Through the mixing process, the sidebands are formed at the sum and the difference of the input frequencies, see figure 5.4. Assuming up-conversion in an ideal mixer:

$$\varphi_{RF,USB} = \varphi_{LO} + \varphi_{VCO-IF}, \tag{5.3a}$$

$$\varphi_{RF,LSB} = \varphi_{LO} - \varphi_{VCO-IF}, \tag{5.3b}$$

for the upper and the lower sideband, respectively. In the phase-locked loop, most computations, e.g. the response to a modulation by noise, involve changes from steady state and are evaluated at an offset from the center frequency. The center frequency itself (shifted by mixing) is not relevant [Ega00]. For this analysis, we can set the phase of the local oscillator $\varphi_{LO} = 0$, so there is no need to explicitly regard this phase offset. If the PLL acts on a mixing product, the phase-frequency detector **must compare** the sideband phase to the reference phase[9] φ_{ref}, and in locked state we obtain:

$$\varphi_{ref} \stackrel{!}{=} \varphi_{RF} = \varphi_{LO} \pm \varphi_{VCO-IF} = \pm \varphi_{VCO-IF}. \tag{5.4}$$

This approach is justified as long as it does not jeopardize loop stability, i.e. as long as we design the loop for sufficient phase margin. The phase margin is quantified as the difference to $180°$ of the phase of the $H(s)$ at a frequency offset where the magnitude of $H(s)$ equals unity[10]. In the following, the lower sideband is selected in order to keep the operating frequency of the PLL as small as possible. When choosing the lower sideband to be routed back into the loop, care must be taken that the phase reversal of φ_{VCO-IF} is taken into account. The mixer's conversion loss (negative gain) can be represented in the respective transfer function by a factor K_m. Otherwise, it can be assumed that an ideal mixer does not affect the characteristics of the loop [Gar79]. The use of a real device, however, implies two effects that occur in the TL: (a) the noise figure of the mixer degrades the SNR; (b) a convolution of the phase power spectral densities of the signals at the mixer's IF and LO ports. These effects will be discussed in section 5.5.

5.2.2.2 Low-pass equivalent for band-pass filtering

It is crucial to determine the effect of the band-pass filter on the loop transfer functions, as it will eventually affect the loop response to phase noise. Gardner and later, Egan, have shown that it is in general possible to find a low-pass equivalent representation for a band-pass filter in a PLL as long as the modulation deviation is small[11] ([Gar79], [Ega99]).

The low-pass - band-pass transformation is a technique often employed in filter

---

[9]In the case where no dividers are used.

[10]The subtraction at the phase detector gives an additional $180°$, and the condition for constant amplitude oscillation is a gain of 1 at a round-trip phase of $360°$.

[11]The higher-order FM sidebands are not (necessarily) represented correctly by this approach, so they should be small for it to be valid [Ega99].

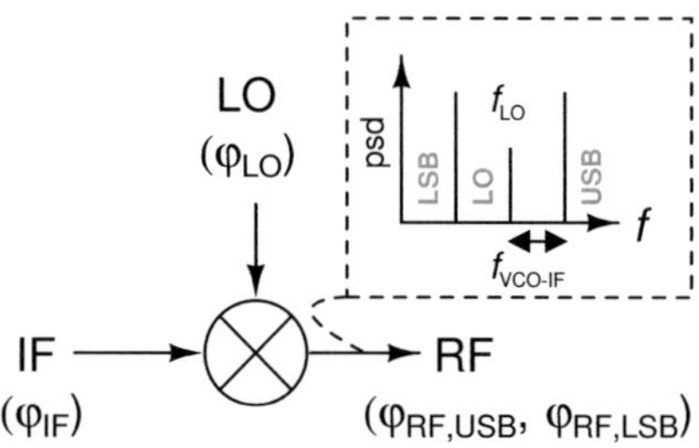

Figure 5.4: Ideal unbalanced mixer

design, whereby a low-pass prototype is modeled according to the desired characteristics, e.g. with a maximally flat or an equiripple response, and subsequently transformed into a band-pass at the desired center frequency, which only requires a scaling in frequency ([MYJ64], [KK06]): for a symmetrical band-pass,

$$s = \frac{(s')^2 + (2\pi)^2 f_{p1} f_{p2}}{2\pi(f_{p2} - f_{p1})s'}, \tag{5.5}$$

where s is the Laplace variable in the normalized low-pass domain, s' is the Laplace variable in the desired band-pass domain, and f_{p1} and f_{p2} are the passband edges of the filter. A symmetrical band-pass is in this context one for which the relative widths from stopband edge to passband edge are equal on both sides of the filter

$$\frac{f_{d1}}{f_{p1}} = \frac{f_{d2}}{f_{p2}}, \tag{5.6}$$

where f_{d1} and f_{d2} are the stopband edges[12]. In principle, the inverse transformation can be used to determine the low-pass equivalent of a given band-pass filter from its measured characteristic, e.g. from a commercial filter. In our case, the actual situation is even simpler as we design the filter ourselves and start out from the low-pass prototype. The low-pass prototype is designed to follow an attenu-

---

[12]For unsymmetrical band-passes, where the widths from passband to stopband edge are equal in absolute values, the Zdunek transformation can be used [KK06].

ation characteristic of $G^2_{\text{eqLPF}}(f)$. In terms of the filter transfer function Z_{eqLPF}, $G^2_{\text{eqLPF}}(f) = |Z_{\text{eqLPF}}(jf)|^2$. The poles of Z_{eqLPF} are the additional poles that will appear in the overall transfer function of the modified PLL.

5.2.2.3 Amplifiers in the loop

An amplifier in the loop can, in a first approximation, be represented by a gain factor K_a. In a phase-locked loop optimized for CW operation, the amplifiers operate at a fixed frequency, which is why the gain can be thought of as constant.

5.3 PLL system demonstrator

The architecture shown in figure 5.2 can be represented in an equivalent electrical circuit depicted in figure 5.5. The dashed box labeled "LO" in the upper right corner represents the up-conversion stage, i.e. the optical up-conversion sub-system. The goal is to stabilize the sideband at a frequency f_{RF} corresponding to one of the sidebands which result from the mixing process of the mm-wave carrier signal supplied by the MLLD (at $f_{\text{LO}} = \nu_{\text{F}}$) and the modulation signal at a center frequency f_{IF}. Ideally, f_{RF} corresponds to the center frequency of the spectral mask specified in the standard documents. For the stabilization concept to operate correctly, we **require** that (refer to figure 5.5) in locked state the reference phase (divided down by a ratio R) is equal to the (scaled-down) phase of the sideband phase

$$\frac{\varphi_{\text{ref}}}{R} \overset{!}{=} \frac{\varphi_{\text{LO}} \pm \varphi'_{\text{IF}}}{N}.$$

Here, φ_{LO} is the phase of the mm-wave carrier, and φ'_{IF} is the phase of the IF signal into the LO stage. The loop, however, does not actually act on the same sideband signal, but rather on the one that is re-created at the output of the loop mixer. In locked state, its phase is $\varphi'_{\text{LO}} \pm \varphi_{\text{IF}}$ and it is

$$\frac{\varphi_{\text{ref}}}{R} = \frac{\varphi'_{\text{LO}} \pm \varphi_{\text{IF}}}{N},$$

where φ'_{LO} is the phase of the *recovered* carrier signal, and φ_{IF} is the phase of the IF signal out of the IF oscillator. φ_{IF} and φ'_{IF} differ only insofar as the IF signal experiences a non-negligible time delay through the I/Q modulation stage (on its way to the LO stage). φ'_{LO} and φ_{LO} differ if the carrier signal experiences a non-negligible time delay in the carrier recovery stage. For the following considerations, we will first assume that $\varphi'_{IF} = \varphi_{IF}$ and $\varphi'_{LO} = \varphi_{LO}$.

For a proof of concept, or a performance evaluation of the demonstrator, electrical measurements are sufficient. In an all-electrical system, a signal generator and a mixer circuit are used in the up-conversion stage. While the LO signal is always present in an optically generated DSB-IM spectrum, electrical mixers are usually designed to suppress the LO component in the RF output spectrum. In our case, the mixer actually needs to have significantly *low* LO-to-RF isolation for the DSB to exhibit a strong component at the LO frequency so that it can easily be recovered by band-pass filtering (and additional amplification stages). By first testing the demonstrator in a conventional electrical system, we will be able to estimate the improvement to be expected from further exploiting best practice techniques in PLL design.

The demonstrator is then tested in the RoF environment using two different optical sub-systems. One is the MLLD-based system: when the MLLD-PD pair is employed for LO generation and up-conversion, we implicitly deal with two aspects, namely, the frequency jitter inherent to the MLLD, and its actual phase noise. We have no means to separate these two. The performance of the MLLD-PD pair can however be emulated by means of another system based on a highly stable single-mode DFB laser which we can deliberately perturb by phase noise. A detailed description of the technique will be given in section 5.6.1.

Figure 5.5 corresponds to the actual implementation of the demonstrator: the loop includes the loop mixer, a band-pass filter BPF_1 and an amplifier (characterized by its gain K_{a1}) which compensates for the mixer conversion loss K_m and the insertion loss of BPF_1. The lower sideband is selected in order to keep the N divider ratio as small as possible. The carrier recovery stage is made of the filter BPF_2 which extracts the LO signal from the double sideband spectrum. Two additional amplifiers (characterized by their gains K_{a2}, K_{a3}) have to be included in order to provide the necessary drive power for the loop mixer.

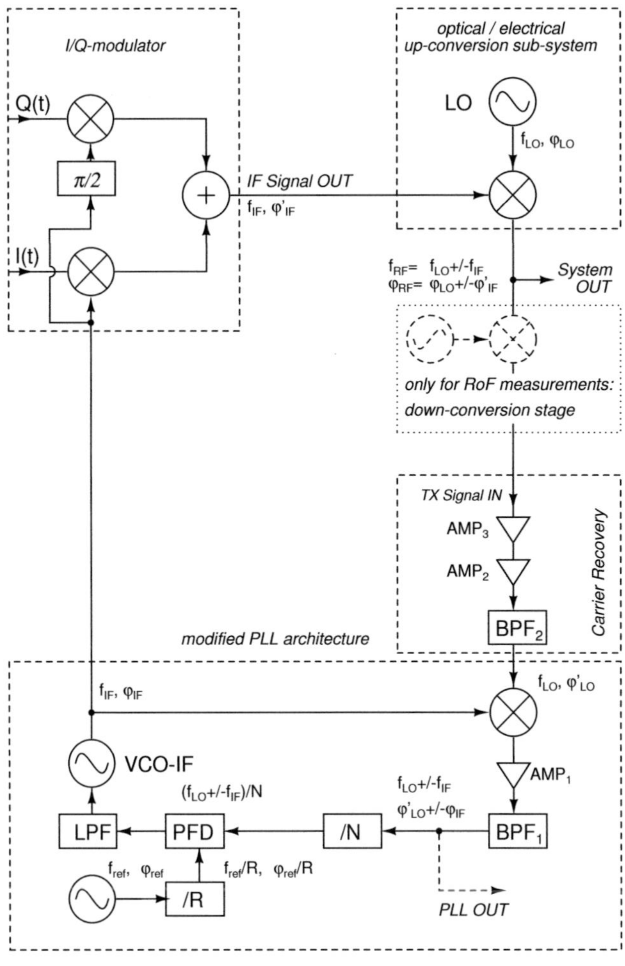

Figure 5.5: Sideband stabilization technique for the RoF system, electrical equivalent as prepared.

The demonstrator features two outputs for the measurement: the first, labeled "PLL OUT", allows us to measure the closed-loop PLL performance directly after BPF_1 (sideband with a phase $\varphi'_{LO} \pm \varphi_{IF}$). Measurements performed at this output allow us to determine the loop response of the PLL itself without further influence of the optical system, *except* MLLD-induced phase noise. The second output, labeled "System OUT", allows us to measure the PLL performance when operating on the single-tone sideband in the full communication system (sideband with a phase $\varphi_{LO} \pm \varphi'_{IF}$). In the case of the all-electrical measurements, both configurations give similar results if identical mixers are used. When the optical sub-systems are used, the two configurations reflect the degrading effects of the (additive) noise added by the optical link, as quantified by the system noise figure NF_{total}. Note that a down-conversion stage (dotted box) has to be included for the mm-wave RoF measurements (see section 5.6).

5.3.1 Forward and reverse transfer functions

The transfer functions given in equations (5.2a) and (5.2b) are modified according to the additional components. In analogy to the RoF system measurements, the signal is tapped after the mixer. All components thereafter are considered part of the reverse transfer function, including BPF_1, amplifier and N divider. The mixer becomes a part of the forward transfer function. The factor L_{WD} reflects the fact that only half of the VCO power is routed back to the loop; a Wilkinson divider with a nominal split of 3 dB (i.e. $L_{WD} = 0.5$) is employed at the branching point. The factor (-1) represents the phase reversal in the lower sideband which is compensated for by reversing the polarity of the charge pump. The new forward and reverse transfer functions are then:

$$G_F(s) = (-1) \cdot K_P \cdot Z_{LPF}(s) \cdot \frac{K_V}{s} \cdot L_{WD} \cdot (-1) \cdot K_m, \qquad (5.7a)$$

$$G_R(s) = K_{a1} \cdot Z_{eqLPF}(s) \cdot \frac{1}{N}. \qquad (5.7b)$$

Qualitatively, the transfer functions are the same when measuring at the system output, except for the insertion loss of the I/Q modulator and the modification of the LO signal by the amplifiers and BPF_2. Quantitatively, the loop response differs considerably depending on which type of LO generation is used.

5.3.2 Conditions for capture and hold-in

In the proposed architecture, the PLL is supposed to lock the modulation sideband generated at the loop mixer from the output signal of the IF oscillator and the recovered LO signal. The *capture* range refers to the range of frequencies for which the unlocked PLL can become locked to the reference after a certain capture time T_{cap}[13]. Once locked, it is required to counter-act the phase and frequency deviations of the modulation sideband which are in particular influenced by the variation of LO frequency and phase. The range of frequencies for which the PLL can track the variations of the controlled signal is referred to as *hold-in* range. Textbooks on PLL theory usually treat the case of capture and hold-in within the context of different excitations - such as phase or frequency steps, or a frequency ramp - at the *reference input* of the PLL ([Ega00], [Bes07]). For applications, where the PLL acts as a frequency synthesizer, or a demodulator, this is a purposeful approach.

In our context, however, the reference is always fixed in frequency and phase. The perturbation we deal with are the phase and frequency deviations of the modulation sideband to be controlled after the loop mixer. These deviations depend on both the IF oscillator and the LO used in the up-conversion stage. We consider the following cases - both of which will be reproduced in the RoF system measurements (see section 5.6):

- **Case 1:** The LO center frequency is stable; the LO exhibits sufficient long-term stability. The variation of the modulation sideband frequency is mostly influenced by the IF oscillator. Through the mixing process at the loop mixer, i.e. the subtraction of input phases, the phase of the modulation sideband experiences a sudden shift (step). This situation corresponds to the DFB-based RoF system.

- **Case 2:** The LO center frequency is not stable; it exhibits only limited long-term stability, as in the MLLD-based RoF system. At the loop mixer,

---

[13] Some textbooks distinguish between the lock-range (lock is obtained within the subsequent cycle) and the pull-in range (lock is achieved only after a certain time span) [Bes07].

a slow variation of the sideband frequency will occur which is due to the variations of both the IF and the LO oscillator frequency. For the MLLD-based system, the frequency jitter of the MLLD dominates over the variations introduced by the IF oscillator. At the loop mixer, both the frequency and the phase of the modulation sideband experience a sudden change.

For the following, we assume that the PLL comprises a charge-pump PFD (see also figure 5.2). In locked state, a PLL featuring a charge-pump PFD does not have a residual frequency error, nor a steady-state phase error [Bes07].

5.3.2.1 PLL in unlocked state: Capture range

The question of whether or not the PLL will lock is determined by two aspects. The first is sufficient phase margin. While a very large phase shift might be produced at the mixer, it primarily concerns the center frequency of the signal to be controlled (here: $f_{\mathrm{LO}} - f_{\mathrm{IF}}$). It is important to note that this phase shift at the **absolute** center frequency does not enter into the loop model. The phase margin is calculated at the unity gain **offset modulation frequency** $f_{\mathrm{G=1}}$ from the center frequency of the controlled signal. The notion of phase margin is essentially independent from the absolute frequency or phase of the controlled signal and can be considered separately in loop design.

The second aspect is the actual capture range of the PLL. If a PFD is used, the average output signal of the PLL varies monotonically with the frequency error. Theoretically, the capture range is infinite[14] [Bes07]. In practice, it is of course limited by the frequency range of the controlled oscillator. As a consequence, the PLL will always lock - if the phase margin condition is met - and the frequency error will settle to zero. Likewise, the range of possible phase error allowed for stable operation is the full $360°$ range for a PFD [Ega00]. Any phase error introduced at the loop mixer can therefore be compensated for, i.e. the residual phase will also settle to zero.

---

[14]The reason is that the loop filter either comprises a real integrator (a capacitance C_1), or behaves like an integrator when driven from a charge pump PFD. The loop's DC gain hence becomes infinite and the scaled-down signal frequency will be pulled up or down until it comes close to the reference frequency.

For case 1, we can therefore be sure that the PLL will lock. For case 2, the situation is more complicated. As T_{cap} is not infinitesimally short, the change in LO frequency must be slow compared to the time span necessary for capture, assuring that the PLL can actually follow the frequency variation of the sideband. In general, T_{cap} depends on several parameters [Bes07],

$$T_{cap} = 2 \cdot \Delta f_0 \cdot \frac{NC_1}{K_P K_V},$$

(5.8)

where Δf_0 represents the difference between the reference frequency and the scaled-down initial frequency of the controlled signal (here, the modulation sideband), and C_1 is the loop filter capacitance that loads the PFD. In equation (5.8), we have neglected the gains of the loop amplifier, the band-pass filter, and the loop mixer. All values considered, T_{cap} can be estimated in the range of hundreds of ns to 1 μs for the frequency offsets produced by the MLLD-based RoF system. We have stated before that the frequency variation of the MLLD beat note is in the order of a couple of split seconds. The capture time is therefore sufficiently short for the PLL to lock.

5.3.2.2 PLL in locked state: Hold-in range

If the PLL is locked, it is the hold-in range which is of interest. The hold-in range is obtained by calculating the frequency where the phase error is at its maximum. In case 1, the use of a charge-pump PFD results in a hold-in range which is theoretically infinite [Bes07] as any phase error can be corrected by the PLL. As for the real capture range, the real hold-in range is again limited by the operating range of the controlled oscillator.

In case 2, as before, the charge-pump PFD will correct the frequency error of the locked PLL. As the LO frequency changes over time, two conditions must however be fulfilled in order to exploit the hold-in range of the PLL. First, the LO frequency must not reach a value that would result in a difference frequency outside the IF oscillator's frequency range. We have previously seen that the MLLD frequency varies within a range of about 600 kHz from its nominal value while even narrow-band VCOs cover ranges of several MHz. Thus this first condition is most certainly valid. Second, the control mechanism of the PLL must be fast enough to follow the frequency variations. Here the decisive factor is the loop bandwidth. It can easily be designed to cover several hundreds of kHz to several MHz, and the frequency variation of the MLLD-generated LO is slow ($\approx$ ms) compare to the reciprocal of the loop bandwidth ($\approx$ 5 µs). The second condition is therefore also fulfilled.

We thus expect the PLL to lock and stay locked, even if the LO frequency varies within a range of as much as 600 kHz.

5.3.2.3 Time delays in the stabilization architecture

So far we have neglected the possible occurrence of time delays in the stabilization architecture. In particular, we have assumed that $\varphi'_{\mathrm{IF}} = \varphi_{\mathrm{IF}}$ and $\varphi'_{\mathrm{LO}} = \varphi_{\mathrm{LO}}$, i.e. no significant time delays occur either for the IF signal on its way from the IF oscillator to the up-conversion stage, or for the carrier signal when it is recovered from the detected DSB spectrum. In the event where a time delay is included at some point in the architecture, we might run into problems regarding loop stability or coherence.

Figure 5.6 shows the potential spots where a time delay might occur in the architecture: T_{d1} represents a delay between the IF oscillator and the up-conversion stage. T_{d2} is the time delay introduced by the carrier recovery stage (including, if applicable, O/E conversion in figure 5.2). Finally, T_{d3} corresponds to a time delay within the loop.

We have previously assumed that

$$\varphi'_{\mathrm{IF}}(t) = \varphi_{\mathrm{IF}}(t) - 2\pi f_{\mathrm{IF}} T_{\mathrm{d1}} \approx \varphi_{\mathrm{IF}}(t), \tag{5.9}$$

$$\varphi'_{\mathrm{LO}}(t) = \varphi_{\mathrm{LO}}(t) - 2\pi f_{\mathrm{LO}} T_{\mathrm{d2}} \approx \varphi_{\mathrm{LO}}(t). \tag{5.10}$$

For $T_{\mathrm{d1}}, T_{\mathrm{d2}} \approx 0$, the sideband phase of the transmit signal is locked to the reference phase which is clearly the desired function of the PLL. If T_{d1} cannot be neglected, the phase of the sideband generated inside the loop is locked to the reference phase, but the phase of the sideband to be transmitted is not. This im-

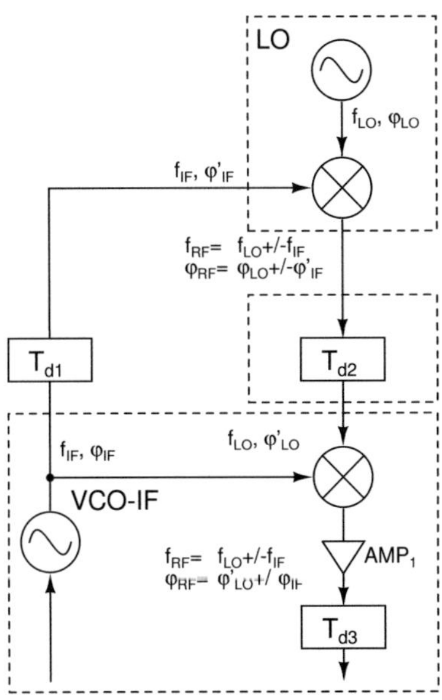

Figure 5.6: Time delays occurring at different spots in the stabilization architecture

plies that there will be a phase offset between the stabilized sideband in the loop and the transmitted sideband - which represents in itself a loss of coherence. The phase offset $(-2\pi f_{IF}T_{d1})$ is deterministic by nature, and it is not related to the loop model. It therefore does not affect the stabilization mechanism.

If T_{d2} cannot be neglected, the carrier recovery stage represents a deterministic offset from the instantaneous phase $(-2\pi f_{IF}T_{d2})$ of the carrier oscillator $\varphi_{LO}(t)$. The additional phase offset of the modulation sideband center frequency can be treated like in the preceding discussion on the phase shift introduced through mixing (case 1). It does not compromise phase margin.

If however, a time delay T_{d3} occurs *within* the loop, the phase shift does not depend on the phase difference of the IF and the LO signal, mixed at the loop mixer. Instead, phase shifts proportional to the respective noise modulation frequency will be produced by T_{d3}. The closed-loop transfer function $H(s)$ is thus affected and modified:

$$H'(s) = H(s) \cdot \exp(-sT_{d3}). \tag{5.11}$$

The phase margin at $f_{G=1}$ is reduced by a value of $(-2\pi f_{G=1}T_{d3})$. If a delay is present in the loop, it is thus wide loop bandwidth which worsens the situation. In general, it is assumed that a delay is not critical as long as it is much smaller than the reciprocal of the loop bandwidth [Ega00]. A conventional PLL is ususally designed for a phase margin[15] of 45°. For a worst case estimation, $f_{G=1}$ approximates the loop bandwidth; and for a loop bandwidth of 200 kHz (as used for the demonstrator) this yields a maximum time delay T_{d3} of 0.625 µs.

Due to the inclusion of the optical sub-system (refer to figure 5.2), certain time delays T_{d1}, T_{d2} on the IF or LO paths might not always be avoided. We have seen, however, that they do not hinder the correct function of the stabilization architecture. Loop stability is only affected if such delays occur within the loop. However, there is no reason why the loop itself should exhibit a non-negligible delay T_{d3}. Compared to the modular approach which was taken in building the system demonstrator, further reductions can be envisioned in the lengths of the signal paths in the PLL by (partly) integrating the components on one board.

5.4 Hardware implementation

In this section, the components of the system demonstrator as prepared will be presented. The demonstrator operates in the μ-wave rather than in the mm-wave range for the simple reason that components are readily available[16] at frequen-

---

[15] The phase margin in the demonstrator is only 27°. This yields a maximum time delay T_{d3} of only 0.375 µs.

[16] VCO-IF: Crystek CVCO55BE series; I/Q modulator: Analog Devices ADL5372; amplifiers: Mini-circuits (1) ZX60-362LN+, (2) ZX60-542LN+, (3) ZX60-V81+; mixers: Hittite HMC128G8; PFD/divider: Analog Devices ADF4108

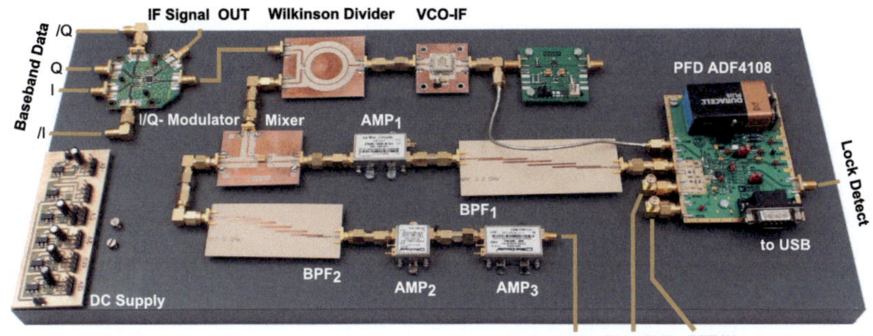

Figure 5.7: Demonstrator hardware. Optional filter PCB not connected. Not shown in the picture: up-conversion stage (IF signal IN, Tx signal OUT)

cies < 8 GHz at the time of writing. The operating frequencies were selected according to the availability of the components. The chosen intermediate frequency is $f_{\text{IF}} = 1.8$ GHz. We assert that the **principle** of sideband stabilization can obviously be demonstrated at μ-wave frequencies. However, for the measurements carried out on the RoF system operating at 60 GHz, the performance of the commercially available PFD chips is not sufficient. We can solve this problem by including the down-conversion mixer that was applied already in the system setup described in figure 3.4 between the up-conversion stage and the carrier recovery stage; it is indicated in figure 5.5 (dashed line components). The carrier frequency is down-converted to $f_{\text{LO}} = 5$ GHz, and the carrier recovery stage was designed for this center frequency. From this choice of frequencies, the resulting sideband frequencies are $f_{\text{RF}} = f_{\text{LO}} - f_{\text{IF}} = 3.2$ GHz and $f_{\text{RF}} = f_{\text{LO}} + f_{\text{IF}} = 6.8$ GHz. A picture of the system demonstrator is shown in figure 5.7. The demonstrator as implemented features the following components.

5.4.1 Loop components

VCO-IF. The surface-mount VCO-IF delivers signals at frequencies from 1.77 to 1.83 GHz at an output power of 3 dBm. Its phase noise at 10 kHz is -106.6 dBc/Hz, and for offset frequencies > 1 MHz, it is 161 dBc/Hz. Its tuning sensitivity is K_V = 25 MHz/V. With a DC voltage of about 2.07 V required for 1.8 GHz, the tuning voltage is in principle low enough to allow for the use of a passive loop filter. Its input capacitance of 47 pF represents a non-negligible capacitive load to the filter circuit and must therefore be taken into consideration during the design of the loop filter. The VCO signal is then split using a non-ideal Wilkinson divider and routed to the loop mixer (split loss of $L_{WD,l,dB}$ = -3.8 dB) and to the I/Q modulator ($L_{WD,r,dB}$ = 4 dB).

I/Q modulator. The surface-mount I/Q-modulator is a vector modulator operating in the range from 1.5 to 2.5 GHz with a maximum output power of 7.1 dBm, an output IP3 of 27 dBm and an output noise floor of -158 dBm/Hz. Its insertion loss is $K_{IQ,dB}$ = -10 dB.

Loop mixer. The mixer in the loop is a surface-mount double-balanced MMIC mixer nominally operating in the range from 1.8 to 5 GHz, with a conversion loss of $K_{m,dB}$ = -8.5 dB. A DSB spectrum is created with principal spectral components at 3.2 GHz (lower sideband), 5 GHz (carrier), and 6.8 GHz (upper sideband); the LO-to-RF suppression of about 30 dB is enough for both the carrier recovery path (where we actually need the LO component in the spectrum) and in-loop up-conversion (where it can be filtered). The additional phase shift of -180° in the lower sideband can be compensated for by reversing the charge pump current of the PFD.

Loop amplifier AMP$_1$. The low noise amplifier in the loop is employed to compensate for the conversion loss of the mixer and the loss of the filter. It is packaged with coaxial connectors and has a gain of $K_{a1,dB}$ = 12 dB and a noise figure of NF_{a1} = 0.9 dB.

Band-pass filter BPF$_1$. The band-pass filter is designed in the low-pass domain with a maximally flat Butterworth response of the following characteristics: passband edge f_c = 50 MHz, stopband edge 100 MHz, passband attenuation L_{ar} = 0.5 dB at f_c, stopband attenuation L_A at the stopband edge -40 dB. From the tabulated Butterworth filter characteristics (see for example [MYJ64]), we find that a filter order of $N = 5$ is sufficient for the desired L_A. The frequency-normalized (i.e. with a cut-off frequency of $\overline{f}_c = 1$ Hz, marked by an overbar) Butterworth squared magnitude response is

$$\overline{G}^2(f) = \left|\overline{Z}_{\text{eqLPF}}(jf)\right|^2 = \frac{G_0^2}{1 + (j2\pi f)^{2N}} \tag{5.12}$$

with a DC gain of G_0 (ideally $= 1$). Since $\overline{Z}_{\text{eqLPF}}(s) \cdot \overline{Z}_{\text{eqLPF}}(-s)$ evaluated at $s = j2\pi f$ is equal to $\left|\overline{Z}_{\text{eqLPF}}(jf)\right|^2$, we can write

$$\overline{Z}_{\text{eqLPF}}(s) \cdot \overline{Z}_{\text{eqLPF}}(-s) = \frac{G_0^2}{1 + (-s^2)^N} \cdot \tag{5.13}$$

The poles of the normalized transfer function of the Butterworth filter can be determined by setting the denominator in equation (5.13) to zero and solving for s. The transfer function is then

$$\overline{Z}_{\text{eqLPF}}(s) = \frac{G_0}{\prod_{k=1}^{N}(s - s_{\text{pk}})}, \tag{5.14}$$

where the poles are

$$s_{\text{pk}} = \exp\left(j(2k + N - 1) \cdot \frac{\pi}{2N}\right), \tag{5.15}$$

$$s_{\text{p1}} = -0.309 + j0.9511,$$
$$s_{\text{p2}} = -0.809 + j0.5878,$$
$$s_{\text{p3}} = -1,$$
$$s_{\text{p4}} = s_{\text{p2}}^* = -0.809 - j0.5878,$$
$$s_{\text{p5}} = s_{\text{p1}}^* = -0.309 - j0.9511.$$

Note that there is only one real pole and two pairs of complex poles. In order to determine the actual transfer function, equation (5.14) must be scaled to a cut-off frequency of $\omega_c = 2\pi f_c = 100\pi$ MHz. The resulting transfer function is

$$Z_{\text{eqLPF}}(s) = \frac{G_0 \omega_c^5}{a_5 s^5 + a_4 s^4 + a_3 s^3 + a_2 s^2 + a_1 s + a_0}. \tag{5.16}$$

The coefficients of the denominator polynomial are tabulated. With frequency-scaling they become [Lac10]:

$$a_5 = 1,$$

$$a_4 = 3.236068 \cdot \frac{1}{\omega_c}, a_3 = 5.236068 \cdot \frac{1}{\omega_c^2},$$

$$a_2 = 5.236068 \cdot \frac{1}{\omega_c^3}, a_1 = 3.236068 \cdot \frac{1}{\omega_c^4},$$

$$a_0 = 1 \cdot \frac{1}{\omega_c^5}.$$

Equation (5.16) is the equivalent filter function to be integrated into equation (5.7b). For the realization of the filter, the filter is shifted to yield the desired passband edges $f_{\text{p1}} = 3.15$ GHz and $f_{\text{p2}} = 3.25$ GHz using the filter transformation described in equation (5.5). The filter can then be realized as a coupled line band pass filter (CL-BPF) and manufactured in-house on microstrip Rogers 4003 substrate. The magnitude of the filter response is measured and is shown in figures 5.8(a). The measured performance is sufficient for the implementation of the filter in the demonstrator. For the loop simulations, we will use the transfer function according to equation (5.16) and adjust G_0 to fit the measured filter loss at the band-pass center frequency $f_{\text{BPF}} = 3.2$ GHz (-3.4 dB at 3.2 GHz, which corresponds to $G_0 = 0.676$). At 3.2 GHz, the measured phase shift is $\theta_{\text{BPF1}} = 0.65°$ (see figure 5.8(b)). Its influence on phase margin is negligible in the direct vicinity of the center frequency, but phase noise at larger offsets will be shaped by the filter function.

Phase-frequency detector, N divider. The main component of the PLL is a chip comprising a low-noise digital PFD, a charge pump, and programmable dividers

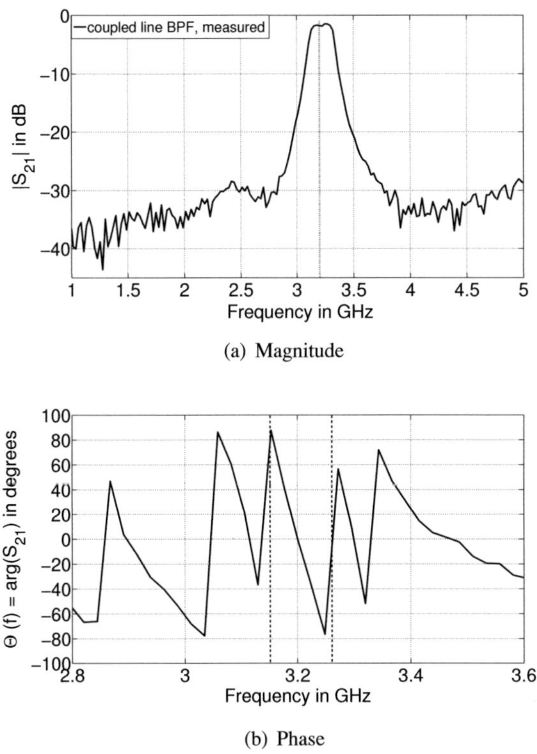

(a) Magnitude

(b) Phase

Figure 5.8: Measured filter response

for both the RF input signal and the reference signal. The use of the digital PFD and charge pump, i.e. the sampling process at the phase detector comparison frequency f_{PFD} associated with these components, poses a problem for the continuous-time description[17]. In the Laplace domain model, the PFD and charge

---

[17]Under certain assumptions, the impulse-invariant transformation allows for the s-domain model to be mapped into the z-domain model in a discrete-time description at a reference time. While the z-transformation gives a precise representation of the effects of sampling, it is inconvenient

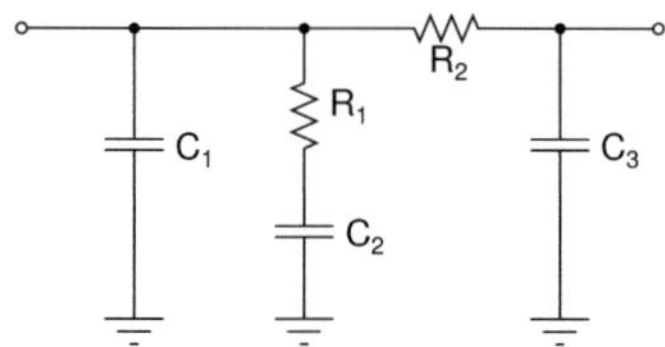

Figure 5.9: Passive 3rd order loop filter comprising one integrator (C_1)

pump block is approximated with a continuous-time block injecting a current proportional to the phase error into the filter. This approximation generally fails to provide the correct transfer function [Gar80]. However, the approximation applies as long as the loop bandwidth is small compared to f_{PFD}; otherwise the PLL becomes unstable. This is the reason why the loop bandwidth is typically designed for no more than 10% of f_{PFD} allowing for the sampling process to be ignored in the transfer function [Ban06]. The maximum charge pump voltage in this case is 5 V, with a maximum charge pump current of $I_P = 7.833$ mA. The resulting charge pump gain is

$$K_P = \frac{I_P}{2\pi} = 1.247 \,\mathrm{mA/rad}. \tag{5.17}$$

The sensitivity of the chip at the input of the N divider is -10 dBm. With $\mathrm{AMP_1}$, an input power of about -6.4 dBm is provided at its input.

Loop filter. With a charge pump voltage of 5 V, and a target control voltage of the VCO-IF of about 2 V, the loop filter can be realized as a passive RC network (refer to figure 5.9) and was designed for third order and a loop bandwidth of 210 kHz at a phase margin of $27°$. The third order passive filter can be accommodated on the same printed circuit board as the commercial PFD. As an option, an external filter PCB for higher filter orders can also be used; in figure 5.7 it is not connected. The components $(C_1 = 5.6\,\mathrm{pF}, C_2 = 220\,\mathrm{pF}, C_3 = 1.5\,\mathrm{pF}, R_1 = 2.7\,\mathrm{k\Omega}, R_2 = 2.7\,\mathrm{k\Omega})$

with respect to building a common transfer function: The z-transform of the composite of two subsequent blocks equals the product of the z-transforms of the individual blocks only if separated by a sampler. Since only one sampler appears in the PLL, the z-transform of the open-loop gain is not the product of the z-transforms of the integrator and the individual components [Ega99].

yield a filter transfer function of

$$Z_{\mathrm{LPF}}(s) = \frac{1 + 5.94 \cdot 10^{-7}s}{2.271 \cdot 10^{-10}s + 5.131 \cdot 10^{-18}s^2 + 1.347 \cdot 10^{-26}s^3}. \qquad (5.18)$$

Simulations have later shown that the large input capacity of the VCO-IF is better driven from an active loop filter at larger comparison frequencies f_{PFD}. The maximum charge pump frequency of the chip is 50 MHz, a loop bandwidth of $f_{\mathrm{BW,max}}$ = 5 MHz can therefore be envisioned. The active loop filter was not implemented in hardware.

Reference signal. The reference signal is delivered by a signal generator[18], allowing the use of different reference frequencies, and selectable comparison frequencies f_{PFD} for the PFD. The reference frequency is set to 200 MHz and the R divider to 100, resulting in a scaling of reference phase noise with a factor of $1/R^2 = 1/10^4$.

5.4.2 Carrier recovery

RoF system measurements: The mm-wave carrier detected by the photo-detector is amplified by the mm-wave amplifier (as in figure 3.4) and down-converted to 5 GHz. It is then split and sent to the spectrum analyzer and the carrier recovery circuit. A carrier power of about -15 dBm at 5 GHz is available for carrier recovery. The carrier recovery stage consists of two amplifiers with gains $K_{\mathrm{a2,dB}}$ = 22 dB (OP1dB = 8 dBm, NF = 1.86 dB) and $K_{\mathrm{a3,dB}}$ = 8 dB (OP1dB 17.5 dBm, NF = 8 dB), as well as the band-pass filter BPF_2 designed as a coupled line bandpass filter centered at 5 GHz. Although it is possible in principle to describe BPF_2 with its transfer function in Laplace notation as we have done before, it will be sufficient to describe the filter by its insertion loss of $K_{\mathrm{BPF2,dB}}$ = -5 dB and the phase shift $\theta_{\mathrm{BPF2}} = 21°$ introduced at 5 GHz. We can do so because its bandwidth of approximately 200 MHz is large compared to the loop bandwidth. The carrier recovery stage was designed to provide at least 10 dBm drive power at the LO port of the loop mixer. With an estimated carrier power of -15 dBm after the down-conversion mixer (i.e. at 5 GHz), a total gain of 25 dB can be achieved at a noise figure of 1.96 dB.

---

[18] Agilent E4433B

Electrical measurements: The carrier signal is provided by a signal generator. The DSB spectrum is created in a mixer identical to the mixer in the loop.

5.4.3 Phase noise measurements on the all-electrical system

The phase noise performance of the loop itself is evaluated using an electrical LO. Here, it is supplied by a signal generator[19] at a frequency of 5 GHz. The signal level is set such that after mixing it with the signal delivered by the VCO-IF in a second mixer (identical to the one described above), the power in the LO component corresponds to the carrier power generated by the RoF system (about -15 dBm, but at a much better SNR). Different levels of LO phase noise can be tested by using Gaussian noise (GN) frequency modulation (FM). This function of the generator requires the specification of a maximum FM deviation Δf. The FM rate is set such that the modulating frequency f_m follows a Gaussian distribution between 0 Hz and 1 MHz, so that the modulation index $\beta = \Delta f / f_m$ is also Gaussian distributed. The modulation appears as phase noise on the generator's CW output signal while the center frequency is fixed. This way, the phase noise characteristics of the LO can be controlled. If we disregard for a moment the fact that two different mechanisms - frequency jitter and phase noise - play a role for MLLD-enabled LO generation, the phase noise level of the signal generator can be adjusted to levels comparable to that of the MLLD in terms of RoF transmission quality as measured by EVM (discussed in chapter 6). The same generator is used as seed generator for the DFB-based RoF system (see section 5.6), where again the phase noise performance can be controlled for a perfectly frequency-stable LO.

Figure 5.10 shows the measured phase psd of the PLL. We distinguish four different regions of interest: at large offsets (I), the noise floor is largely determined by the chip noise scaled by the phase-frequency detector comparison frequency, or the divider ratio N; up to the loop cut-off frequency (210 kHz approx.), the curve

---

[19] Anritsu 68377B

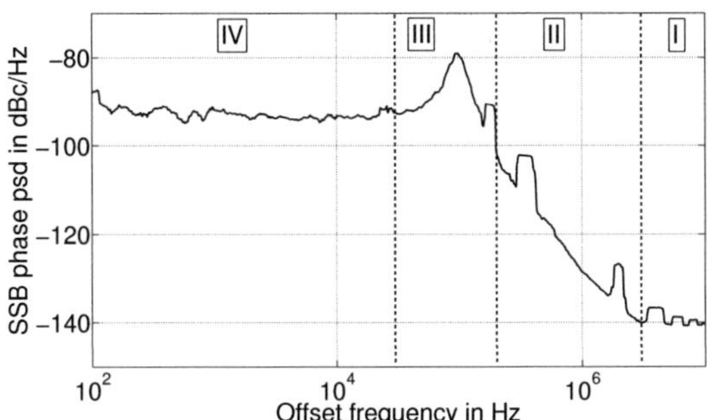

Figure 5.10: Measured phase noise performance in the all-electrical PLL with LO supplied by Anritsu 68377B, no additional GN modulation.

is largely determined by oscillator noise (II); the overshoot can be attributed to the loop filter function (III); and the plateau within the loop bandwidth is, to a large extent, scaled reference noise (IV).

Using this all-electrical setup, the stabilization concept incorporating carrier recovery and the feedback of a filtered sideband could be validated. This constitutes a first-time proof of concept of an operational electrical sideband stabilization and synthesis architecture in the presence of a heavily noisy carrier. The key result of this section and the following section 5.5 have been submitted by the author for publication [BZPC12].

In the following section, the loop response to phase noise shall be broken down into its various components in order to give a more comprehensive picture of the different noise sources that come into play.

5.5 Loop response to phase noise

Phase noise can be understood as the modulation of the carrier frequency by a random noise waveform. In its spectral representation, the noise waveform can be interpreted as a continuum of adjacent passbands of quasi-infinitesimally narrow bandwidth, so that the power in each can be understood as the power of a single sine wave modulating the carrier in phase. It is common practice to treat this phase modulation as a zero-mean stationary random process quantified by its two-sided psd function $S_\varphi(f)$. Its standard deviation, or root-mean-square phase error, is obtained through integration,

$$\sigma_{\text{rms}} = \sqrt{\int_{-\infty}^{+\infty} S_\varphi(f) df}. \tag{5.19}$$

After demodulation, $S_\varphi(f)$ includes both negative and positive frequencies. In phase noise measurements, it is common to consider only the positive frequencies and hence, the single-sideband psd $L_\varphi(f)$. For small[20] phase modulation indices, $L_\varphi(f)$ is obtained from $S_\varphi(f)$ by multiplying with a factor of 0.5 (subtracting 3 dB in logarithmic representation). The phase modulation is considered small if $\sigma_{\text{rms}}^2 \ll 1$ rad^2. In general, it is assumed that this approximation always holds. As we will see in the measurements shown in the following, the mean-square phase error in our case is not significantly smaller than 1 and we cannot actually rely on this approximation. In want of a better approach, we will, however, still use this simple relationship between single-sided and two-sided phase noise in order to at least give an estimate for the rms phase error induced by the system. Practically, the integration is performed within certain limits f_1, f_2. The lower limit f_1 corresponds to a phase noise modulation very close to the carrier. With increasing proximity to zero offset, such phase noise is increasingly difficult to measure. However, it has relatively little impact on the value of the integral (equation 5.19). In practice, we will take f_1 to be the smallest measured frequency. The choice of the upper limit f_2 is more important. With regard to the transmission experiments in the following chapter, we set f_2 such that it covers at least the bandwidth of the signal to be transmitted. In the measurements shown in this chapter, $f_2 = 30$ MHz. The integrated rms phase error which results from

---

[20]The spectrum of a cosine that is sinusoidally phase modulated $\cos(\omega_c t + m \sin(\omega_m t))$ is weighted by the Bessel function J. For small peak phase deviations, $J_0(m) \approx 1$, $J_1(m) \approx m/2$, and the following approximation holds [Ega03]:

$$\cos(\omega_c t + m \sin(\omega_m t)) \approx \cos(\omega_c t) + \frac{m}{2}(\cos((\omega_c + \omega_m)t) + \cos((\omega_c - \omega_m)t)). \tag{5.20}$$

the phase noise measurement on the electrical system shown in figure 5.13 is $\sigma_{\mathrm{rms}} = 1.52°$.

A central element of LTI analysis is the assumption that the response of an oscillator to a set of noise sources, e.g. the different components, is the linear superposition of the responses that would result when only one of the noise sources was applied. The general way to determine the loop's total response is to determine an appropriate closed-loop transfer function H_i from a source i to the output. Starting from the component's power spectral density $S_{\varphi,\mathrm{in},i}(f)$, the resulting power spectral density at the output $S_{\varphi,\mathrm{out},i}(f)$ can be determined according to

$$S_{\varphi,\mathrm{out},i}(f) = S_{\varphi,\mathrm{in},i}(f) \cdot |H_i(jf)|^2 , \tag{5.21}$$

where we have replaced $s = 2\pi f$. Note that the noise waveform is a voltage or a current, so the squared closed-loop transfer function appears in the output power spectral density. The total output psd can then be obtained by summing up the various contributions,

$$S_{\varphi,\mathrm{TOT}}(f) = \sum_i S_{\varphi,\mathrm{out},i}(f). \tag{5.22}$$

This procedure is reflected in the component noise equivalent circuit as depicted in figure 5.11, where each component is assumed ideal and the noise contribution is inserted as a noise source labeled with the respective phase noise psd $S_{\varphi,\mathrm{in},i}$. Note that the noise contribution of the loop filter has been neglected. This is justifiable for passive loop filters; for active loop filters it cannot usually be disregarded. We also neglect the phase noise originating in the N and R dividers, but consider the figure of merit of the PLL chip incorporating these dividers. In order to evaluate the demonstrator loop's response to noise, it is necessary to determine the phase psd of the different components. This will be the subject of the following section.

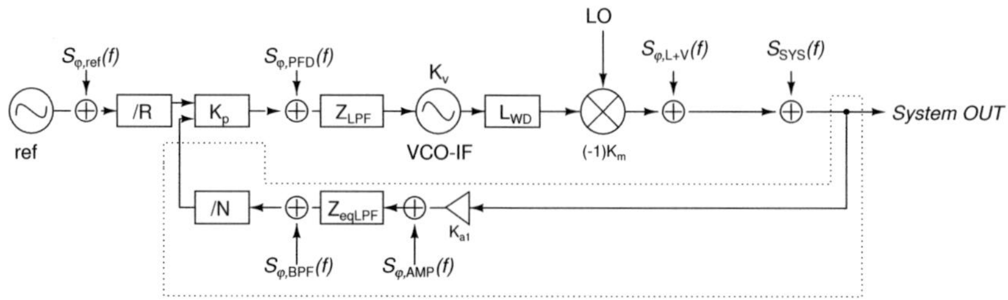

Figure 5.11: Component noise equivalent circuit

5.5.1 Device phase noise models and noise transfer through the loop

Throughout this section, variables are used to describe the various component parameters; a listing of their actual values can be found in table 5.1 at the end of the section. At the input of the loop, the phase noise performance of the reference oscillator sets an absolute limit to the improvement of oscillator phase noise by coupling it to the reference. The reference oscillator used herein is a signal generator operating at 200 MHz. We can describe its phase noise psd $S_{\varphi,\text{ref}}(f)$ in terms of Leeson's model [Lee66], where f_r is the reference frequency, Q_r is the quality factor of the reference, $f_{\text{fc,r}}$ is the flicker corner frequency, and P_r and F_r the delivered power and the noise figure, respectively,

$$S_{\varphi,\text{ref}}(f) = \frac{k_B T F_r}{P_r}\left(1 + \frac{f_{\text{fc,r}}}{f} + \left(\frac{f_r}{2Q_r}\right)^2 \frac{1}{f^2} + \left(\frac{f_r}{2Q_r}\right)^2 \frac{f_{\text{fc,r}}}{f^3}\right). \tag{5.23}$$

Reference phase noise will be scaled down by the reference divider value, and transformed by a low-pass transfer function to the output of the loop:

$$S_{\varphi,\text{out,ref}}(f) = S_{\varphi,\text{ref}}(f) \cdot \frac{1}{R^2}\left|\frac{G_F(jf)}{1 + G_F(jf)G_R(jf)}\right|^2. \tag{5.24}$$

It is therefore preferable to pick a high f_r if the reference phase noise can be scaled down by a high value of R.

The phase noise psd's of electrical components such as dividers, amplifiers, and phase-frequency detectors typically show a flat region as well as a flicker (f^{-1}) region. We will however neglect the flicker region, as it is cumbersome to measure reliably. The phase noise of the phase-frequency detector chip is characterized by the chip's figure of merit (FOM, specified in dB in the datasheet). The actual contribution of the chip scales with the operating frequency of the phase-frequency detector, f_{PFD}. The chip noise is then

$$S_{\varphi,\mathrm{PFD}}(f) = 2 \cdot 10^{(\mathrm{FOM}+10\log f_{\mathrm{PFD}})} \left(1 + \frac{f_{\mathrm{fc,PFD}}}{f} \right) \approx 2 \cdot 10^{(\mathrm{FOM}+10\log f_{\mathrm{PFD}})}, \quad (5.25)$$

where we neglect the influence of the chip flicker noise. The chip noise will be scaled by the charge pump gain and transformed to the output of the loop, again by a low-pass transfer function:

$$S_{\varphi,\mathrm{out,PFD}}(f) = S_{\varphi,\mathrm{PFD}}(f) \cdot \left| \frac{1}{K_{\mathrm{p}}} \right|^2 \left| \frac{G_{\mathrm{F}}(jf)}{1 + G_{\mathrm{F}}(jf)G_{\mathrm{R}}(jf)} \right|^2 . \quad (5.26)$$

The noise introduced by the band-pass and mixer can be estimated, neglecting flicker noise, by means of the component's noise figure, whereby the noise factor F_{bp} of the band-pass filter is not measured directly but estimated from its insertion loss at the band center frequency. The noise introduced by the band-pass in the feedback path can then be approximated as

$$S_{\varphi,\mathrm{BPF}}(f) \approx \frac{k_{\mathrm{B}}T F_{\mathrm{bp}}}{P_{\mathrm{bp}}}, \quad (5.27)$$

where P_{bp} is the power into the filter. This noise will be transformed by the loop according to

$$S_{\varphi,\mathrm{out,BPF}}(f) = S_{\varphi,\mathrm{BPF}}(f) \cdot N^2 \cdot \left| \frac{G_{\mathrm{F}}(jf)}{1 + G_{\mathrm{F}}(jf)G_{\mathrm{R}}(jf)} \right|^2 . \quad (5.28)$$

P_{a1} is the power into the amplifier. The noise introduced by the amplifier with a noise factor F_{a1} in the feedback path can be approximated as

$$S_{\varphi,\mathrm{AMP}}(f) \approx \frac{k_{\mathrm{B}} T F_{\mathrm{a1}}}{P_{\mathrm{a1}}}, \tag{5.29}$$

and this noise will be transformed by the loop according to

$$S_{\varphi,\mathrm{out,AMP}}(f) = S_{\varphi,\mathrm{AMP}}(f) \cdot \left| Z_{\mathrm{eqLPF}}(s) \right|^2 \cdot N^2 \cdot \left| \frac{G_{\mathrm{F}}(jf)}{1 + G_{\mathrm{F}}(jf) G_{\mathrm{R}}(jf)} \right|^2. \tag{5.30}$$

In the electrical experiments, LO phase noise originates directly in the signal generator; additional GN modulation might be employed. For the RoF measurements, LO phase noise is the phase noise of the mm-wave carrier generated in the RoF system *and convoluted* with the phase noise of the down-conversion mixer used to bring the LO signal from the mm-wave range down to 5 GHz. Note that what is considered here as "LO phase noise" is the phase noise originating from the respective source (including down-conversion for the RoF system), *transformed* by the amplifiers (gains K_{a3}, K_{a2}) and the band-pass filter BPF$_2$ (refer to figure 5.12). As we actually measure the resulting phase noise psd after BPF$_2$, and base our derivation of total phase noise at the output of the loop thereupon, we will not explicitly derive a transfer function for BPF$_2$.

Both LO and VCO-IF phase noise can be described according to Leeson's model:

$$S_{\varphi,\mathrm{VCO}}(f) = \frac{k_{\mathrm{B}} T F_{\mathrm{V}}}{P_{\mathrm{V}}} \left(1 + \frac{f_{\mathrm{fc,V}}}{f} + \left(\frac{f_{\mathrm{V}}}{2Q_{\mathrm{V}}} \right)^2 \frac{1}{f^2} + \left(\frac{f_{\mathrm{V}}}{2Q_{\mathrm{V}}} \right)^2 \frac{f_{\mathrm{fc,V}}}{f^3} \right), \tag{5.31}$$

$$S_{\varphi,\mathrm{LO}}(f) = \frac{k_{\mathrm{B}} T F_{\mathrm{LO}}}{P_{\mathrm{LO}}} \left(1 + \frac{f_{\mathrm{fc,LO}}}{f} + \left(\frac{f_{\mathrm{LO}}}{2Q_{\mathrm{LO}}} \right)^2 \frac{1}{f^2} + \left(\frac{f_{\mathrm{LO}}}{2Q_{\mathrm{LO}}} \right)^2 \frac{f_{\mathrm{fc,LO}}}{f^3} \right). \tag{5.32}$$

The VCO signal passes through the power divider and suffers a conversion loss by the loop mixer,

$$S'_{\varphi,\mathrm{VCO}}(f) = \left| L_{\mathrm{WD}} \right|^2 \cdot \left| (-1) \cdot K_{\mathrm{m}} \right|^2 \cdot S_{\varphi,\mathrm{VCO}}(f). \tag{5.33}$$

If we consider the first mixer instead, the resulting phase psd differs from equation (5.33) only by the insertion loss of the I/Q modulator. In the carrier recovery stage, the LO signal is amplified, filtered and attenuated by the mixer's LO-to-RF isolation:

$$S'_{\varphi,\mathrm{LO}}(f) = |K_{a3}|^2 \cdot |K_{a2}|^2 \cdot |K_{BPF2}|^2 \cdot |\exp(j\theta_{\mathrm{BPF2}})|^2 \cdot |K_{\mathrm{LO-RF}}|^2 \cdot S_{\varphi,\mathrm{LO}}(f).$$
(5.34)

From 5.34, we see that by carefully selecting the amplifiers and the filter, it is possible to design the carrier recovery stage such that it is quasi-transparent with respect to LO phase noise. The loop mixer and the mixer in the up-conversion stage can therefore be operated under similar conditions.

When the recovered LO signal and the VCO-IF signal pass through the mixer, the phase psd's at the mixer's RF output will be convoluted to give:

$$S_{\varphi,\mathrm{L+V}} = S'_{\varphi,\mathrm{LO}}(f) * S'_{\varphi,\mathrm{VCO}}(f).$$
(5.35)

Through the loop, this combined phase noise will be transformed by a *high-pass* transfer function,

$$S_{\varphi,\mathrm{out,L+V}} = S_{\varphi,\mathrm{L+V}} \cdot \left| \frac{1}{1 + G_{\mathrm{F}}(jf)G_{\mathrm{R}}(jf)} \right|^2.$$
(5.36)

Additive noise from the RoF system is the most important noise contribution from outside the loop bandwidth. It determines the recovered LO's noise floor such that $k_{\mathrm{B}}TF_{\mathrm{LO}}/P_{\mathrm{LO}} \approx k_{\mathrm{B}}TF_{\mathrm{SYS}}/P_{\mathrm{out}}$ after the mixer, where F_{SYS} can be calculated from the system noise figure, and the component noise factors F_{a2}, F_{a3} and F_{BPF2} according to the Friis equation (see equation (3.37)). Since F_{total} is high and leads the chain, the following approximation can be used:

$$S_{\varphi,\mathrm{SYS}}(f) = \frac{k_{\mathrm{B}}TF_{\mathrm{SYS}}}{P_{\mathrm{in}}} \approx \frac{k_{\mathrm{B}}TF_{\mathrm{total}}}{P_{\mathrm{in}}}.$$
(5.37)

For the electrical measurements, the noise figure of the recovered LO is determined by the signal generator's noise floor. This noise will be transformed through the loop in a high-pass transfer function:

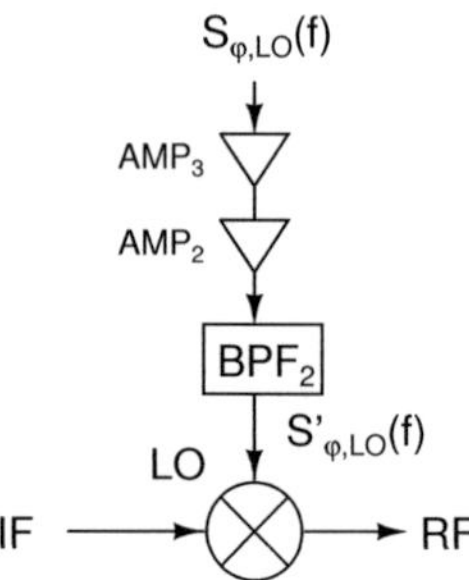

Figure 5.12: Carrier recovery. Detail from figure 5.5

Table 5.1: Loop components and parameter values, electrical measurements. (*) datasheet value, (**) estimate from simulation

Device	Parameters
REF	$P_{\mathrm{r}} = -3$ dBm, $f_{\mathrm{fc,r}} = 1$ kHz, $NF_{\mathrm{r}} = 23$ dB (**), $Q_{\mathrm{r}} = 50000$ (**)
PFD	FOM $= -223$ dBc/Hz (*), $f_{\mathrm{PFD}} = 2$ MHz, N $= 1600$
BPF$_1$	$NF_{\mathrm{bp}} = 3.4$ dB, $P_{\mathrm{bp}} = -3$ dBm
AMP$_1$	$NF_{\mathrm{a1}} = 0.9$ dB, $P_{\mathrm{a1}} = -15$ dBm
VCO-IF	$P_{\mathrm{V}} = 3$ dBm, $f_{\mathrm{VCO-IF}} = 1.8$ GHz, $f_{\mathrm{fc,V}} = 3$ MHz (*), $NF_{\mathrm{V}} = 41$ dB (*), $Q_{\mathrm{V}} = 900$
LO	$P_{\mathrm{LO}} = -15$ dBm, $f_{\mathrm{LO}} = 5$ GHz, $NF_{\mathrm{LO}} = 18.97$ dB, $Q_{\mathrm{LO}} = 625$

$$S_{\varphi,\mathrm{out,SYS}}(f) = S_{\varphi,\mathrm{SYS}}(f) \cdot \left| \frac{1}{1 + G_{\mathrm{F}}(jf)G_{\mathrm{R}}(jf)} \right|^2. \tag{5.38}$$

Outside the loop bandwidth, the noise floor is thus hardly altered by the loop's response function, and its value corresponds in good approximation to the measured noise figures of the recovered LO. For the electrical measurements, the noise floor is about -140 dBc/Hz, which corresponds to a noise figure in the order of 18 - 19 dB, as measured for the signal generator. For the RoF measurements,

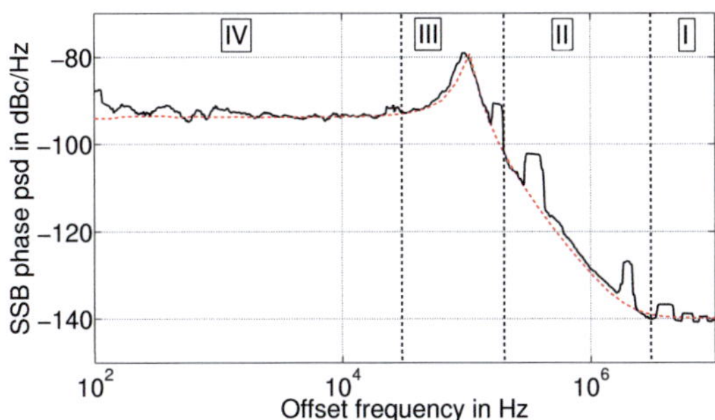

Figure 5.13: Comparison of the simulated (dotted line) and measured (solid line) phase noise performance in the all-electrical PLL with LO supplied by Anritsu 68377B, no additional GN modulation.

the noise floor is raised to -92 dBc/Hz, which corresponds to a noise figure of about 65 dB. This value agrees well with the system noise figure F_{total}, as derived in chapter 3. Table 5.1 lists the values used for the simulations.

We have thus determined all relevant contributions to overall phase noise at the loop output and can calculate the total phase noise psd according to equation (5.22). Based on this analysis, we can now evaluate the improvement to be expected from a more rigorous PLL design.

We have replotted the measurement previously shown in figure 5.10 along with the simulated curve according to the preceding analysis. The result is shown in figure 5.13. It can be observed that the simulation reproduces the measured results very closely. We thus conclude that our analysis of the loop's response to phase noise is valid, and we can use it as a tool to explore the possible improvement of the PLL in the next section.

5.5.2 Design rules for loop optimization in the presence of noise

For the following considerations, it is assumed that the phase noise of the LO is a given parameter which cannot be altered by the system designer. In order to optimize the performance of the loop, the designer should choose a VCO-IF with low phase noise performance. At a given noise level, the optimum loop bandwidth is found at the intersection of the f^{-2} slope and the noise floor. The loop bandwidth sets the minimum phase-frequency detector frequency for stable operation (factor of 10). Inband phase noise can be suppressed by selecting the smallest possible divider ratio N, i.e. the highest possible f_{PFD}. However, outside the loop bandwidth, a high value of f_{PFD} raises the noise floor. For low charge pump supply voltages, high VCO input capacitances and high f_{PFD}, the loop filter is best implemented as an active filter designed for a given value of f_{PFD}, N, loop bandwidth and filter order. The reference signal is crucial for inband phase noise suppression: a high-Q oscillation is provided by a crystal oscillator whose nominal frequency is as high as possible. The reference phase noise can then be scaled down by the R divider.

Adhering to those design rules, the PLL can be improved under the assumption that the LO phase noise and the VCO-IF phase noise stay the same. With the same chip, the PFD frequency can be increased to f_{PFD} = 50 MHz, allowing a minimum divider value of N = 64 for stabilization at 3.2 GHz. The charge pump current was set to I_P = 5 mA. A third order active loop filter in the so-called "standard feedback" topology was simulated incorporating the following components: C_1 = 0.63 pF, R_1 = 6.55 kΩ, C_2 = 19.7 pF, R_3 = 38 Ω, C_3 = 134 pF (Nomenclatura as in [Ban06]). A high-Q oscillator with a noise figure of 23 dB and a factor of Q_r = 1000000, as well as a flicker corner frequency $f_{fc,r}$ = 1 kHz was employed. The improvement of the total phase psd can be observed in figure 5.14, where the integrated phase error can be reduced from 1.52° to about 0.21°.

While the lower limit of integration does not have a severe impact on the accumulated phase error, the upper limit does matter, in particular in wideband systems. At large offsets, the amount of additional phase noise that has to be accounted for when pushing the upper integration limit higher, is highly dependent on system noise, regardless of the inband dynamics of the PLL. This situation is shown

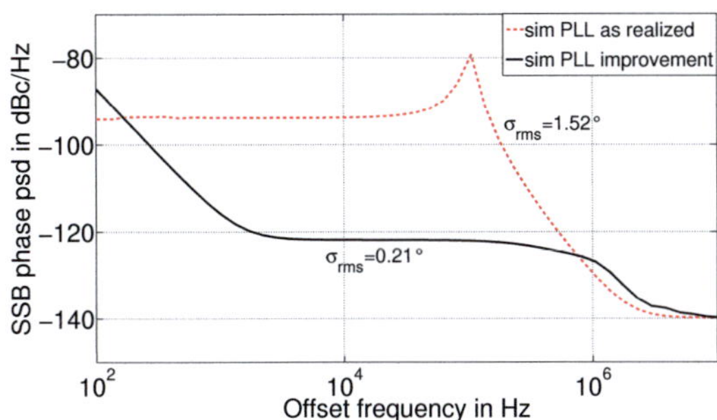

Figure 5.14: Possible σ_{rms} improvement to be expected from PLL design (simulations)

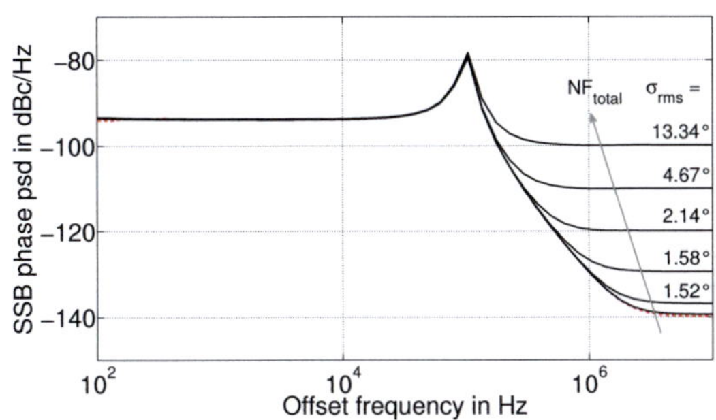

Figure 5.15: Variation of system noise figure (simulations)

in figure 5.15, in which the noise figure was increased in steps of 10 dB up to a value of 70 dB. The values of σ_{rms} resulting therefrom are plotted along with the curves; a variation from 1.52° to 13.34° results. This is why, above all other measures, the system which the loop will operate on must be designed for minimum system noise figure - which for the RoF system implies the highest possible intrinsic link gain. Methods to improve link gain and the system noise figure resulting from this have been presented in chapter 3.

5.6 Phase noise measurements in the stabilized RoF system

RoF measurements were performed on two different RoF transmitter architectures. The first one is the external modulation link which has already been presented in chapter 3 (figure 5.16(a)). The reason why we here prefer the external modulation link to the direct modulation architecture is simply that we can expect a better SNR, and thus, a better performing system from the former. It should be noted, however, that we could also have chosen a direct modulation link for a mere demonstration of the stabilization principle, as its general feasibility was validated in chapters 3 and 4. The second one is a RoF system based on a single-mode DFB laser (figure 5.16(b)), which is equivalent to the MLLD-based system in the sense that it can also deliver an optically up-converted carrier and a modulation in the mm-wave range, a technique presented in [NDCL09]. It is based on a single-mode DFB laser whose output CW signal is directly modulated with the information signal centered at the intermediate frequency. The signal is subsequently up-converted to a carrier frequency of 60 GHz by means of an MZM biased for frequency-doubling, so that the electrical generator operates at f_{seed} = 30 GHz. Frequency-doubling in an MZM is achieved by biasing the MZM at its point of minimum transmission, i.e. in the region of maximum non-linearity. Using the GN modulation option, the phase noise of the generated mm-wave carrier is determined by the seed signal fed into the MZM. Frequency-doubling will lead to a modification of the generator phase noise by $20 \cdot \log_{10}(2)$ = 6.02 dB [Cam98].

In this system, an arbitrary level of phase noise can be emulated while the carrier signal is stable in frequency. In particular, a phase noise level can be generated which is equivalent to the one measured for the carrier in the MLLD-based system. With a maximum frequency deviation of Δf = 80 kHz, a phase noise characteristic similar to that shown in figure 2.14 could be obtained. A comparison is shown in figure 5.17. It will be demonstrated in chapter 6 that both systems

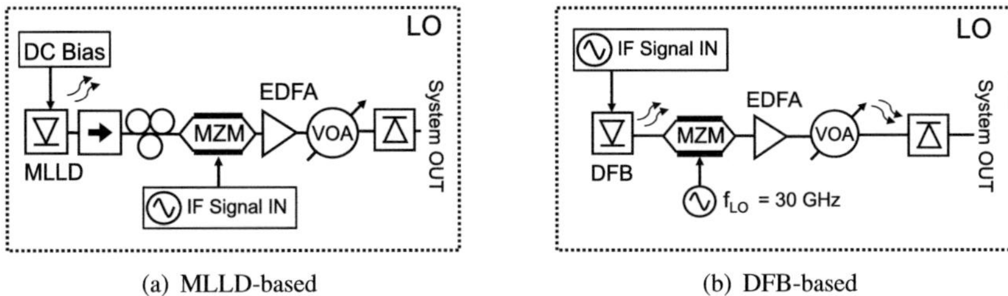

(a) MLLD-based   (b) DFB-based

Figure 5.16: RoF transmitters replacing the up-conversion ("LO") stage in figure 5.5. The "IF Signal IN" box represents a vector signal generator.

yield comparable transmission quality for relatively small signal bandwidths. For large signal bandwidths, the GN capabilities of the electrical generator are not sufficient in order to reproduce the f^{-2} slope at larger offsets. It is interesting to observe that the system noise figure of the DFB-based system is relatively high, although lower than that of the MLLD-based system. The moderate SNR in the DFB-based system can be attributed to the reduced modulation efficiency of the MZM when operated at the minimum transmission point.

In anticipation of the results to be discussed in the following section, we assert that the sideband stabilization concept is also validated for RoF transmission. This is the first time a stabilization and tuning architecture could successfully be demonstrated under the condition of a free-running MLLD; in particular, it is the first demonstration of stabilized RoF transmission in the 60 GHz range under the condition of increased carrier phase noise and frequency jitter. The findings of this section have been submitted by the author for publication [BZP+12].

5.6.1 DFB-based carrier generation

Figure 5.18 shows the performance of the system demonstrator operating on the DFB-based RoF system, both with and without additional GN modulation on the carrier. The center frequency of 60 GHz has been translated to 0 Hz to show the two-sided spectrum around the carrier. Within the loop bandwidth, the PLL

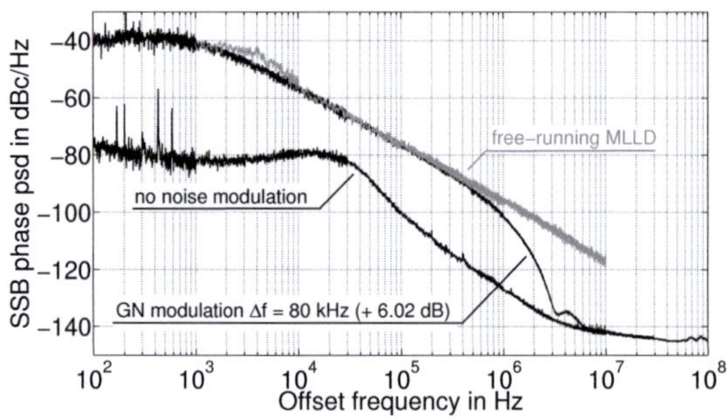

Figure 5.17: Emulation of MLLD phase noise with the DFB-based system, free-running, no PLL.

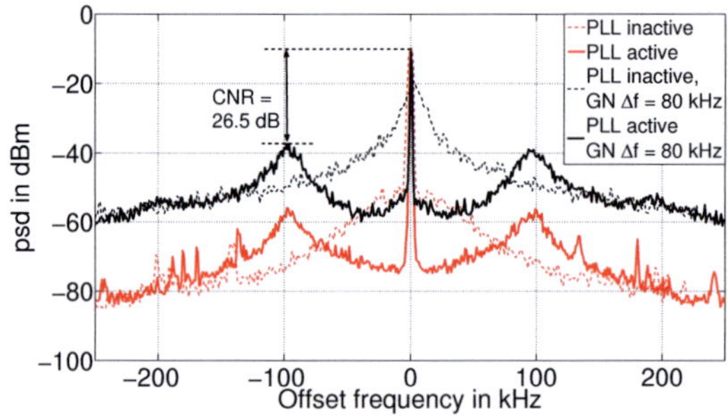

Figure 5.18: DFB-based system, with and without GN modulation ($\Delta f = 80\,\mathrm{kHz}$), RBW = 1 kHz.

improves phase noise in the immediate proximity of the carrier by approximately 46 dB. The effect is more pronounced for GN modulation ($\Delta f = 80\,\mathrm{kHz}$). A carrier-to-noise ratio of 26.5 dB is obtained, measured at the noise humps caused by the overshoot in the closed-loop transfer function.

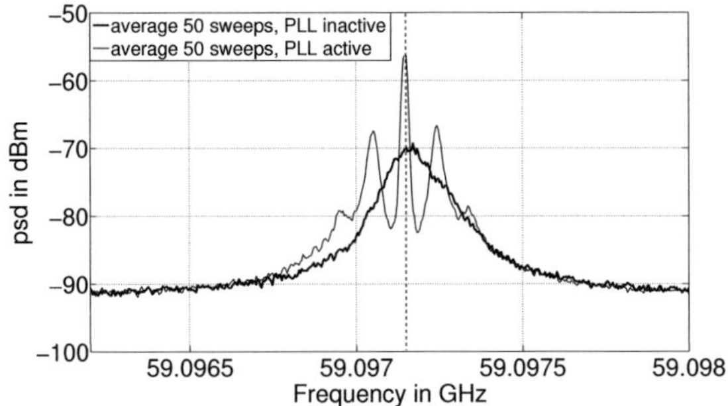

Figure 5.19: MLLD-based system, average of 50 sweeps, measurements with and without PLL (L872), RBW = 10 kHz.

5.6.2 MLLD-based carrier generation

Figure 5.19 shows the measurement of 50 subsequent sweeps (previously shown in figure 2.12) on the MLLD-based system with the PLL in active operation. The measurement was performed on chip L872, the least stable of the three samples. The lower sideband is stabilized at 59.097 GHz ($v_F - f_{IF}$, with v_F =60.897 GHz and f_{IF} = 1.8 GHz). The problem of frequency jitter translating from the carrier to the sidebands can thus be solved. If we want to change the sideband frequency, we can do so by adjusting the operating range of the PLL; in our case, it is defined by the operating range of the VCO and the band-pass filter. Although the implemented hardware does not currently allow for this flexibility, we maintain that the design of a frequency-agile filter in a PLL has been shown in state-of-the-art PLLs.

Figure 5.20 shows the phase noise measurements that have been performed both on the GN modulated DFB-based system (black curves) and on the MLLD-based system (grey curves), with and without the PLL, respectively. In both cases, we observe again that the performance is comparable. The influence of the system can clearly be seen at frequency offsets outside the loop bandwith, where the system noise figure raises the noise floor to about -92 dBc/Hz. The integrated

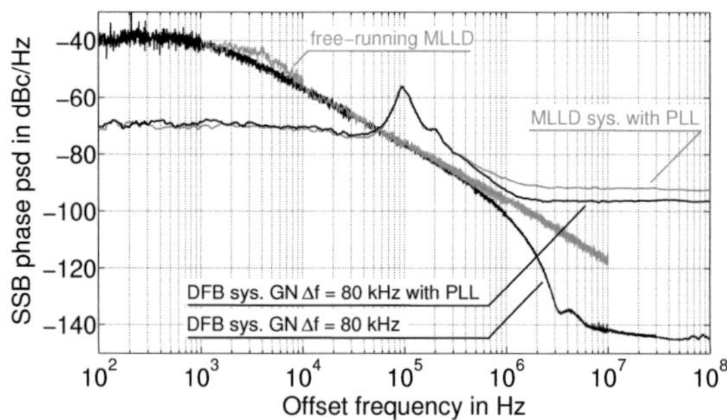

Figure 5.20: Comparison of the DFB- and the MLLD-based systems with and without PLL (L34)

phase error that can be calculated from the MLLD system measurement is $27.53°$. Simulation has shown that the integrated phase error could be improved to $19.80°$ if the modifications to the PLL as indicated in section 5.5.2 were implemented. For further improvement, the noise figure NF_{total} of the system must be reduced.

5.7 Conclusions on chapter 5

In this chapter, we addressed the problem of signal stabilization and frequency tuning. In a communication system in which an information signal is frequency-converted by mixing it with a high frequency carrier, it is the up-converted signal we wish to stabilize or tune. If the signal modulation results in a double sideband spectrum - as in the case of an optical IM system - the stability of the carrier is secondary as long as at least one of the sidebands is stable. Based on this consideration, a conventional PLL (typically used to stabilize and tune a carrier signal) can be modified to operate on a sideband signal. We have referred to this architecture as an up-conversion translation loop. Its most striking feature is that it can provide a stable modulation band in the presence of high carrier instability as manifested in frequency jitter and phase noise.

The loop can therefore be employed to correct the frequency jitter of the mm-wave carrier generated by the MLLD. A first proof of concept was shown based on the system demonstrator also presented in this chapter. Both electrical measurements and RoF measurements on two different systems have shown the general feasibility of the up-conversion translation loop. In the high phase noise DFB-system, the loop performs a considerable phase noise suppression within its loop bandwidth.

In the MLLD-based system, the frequency jitter of the mm-wave carrier can be suppressed entirely. The loop's IF synthesizer capability makes it possible to tune the sideband frequency to a nominal center value as required by the standard spectral masks. By modeling the loop, the oscillators and the noise contribution of the system under test, the measurement results could be reproduced in simulation. By simulation, the requirements on the loop's performance could be refined. If the design rules presented in section 5.5.2 are respected, the resulting rms phase error can be decreased from $27.53°$ to $19.80°$. This constitutes a considerable improvement, but it is still not a satisfactory result with respect to the degradation of transmission quality implied by such high phase errors (see chapter 6).
Further improvement can only be expected if the SNR of the system can be improved. This has become particularly apparent for the RoF system which exhibits a noise figure considerably higher than that of the (generator-based) electrical system. If the system noise figure was decreased to a value comparable to that of the all-electrical system, the rms phase error could further be decreased to $0.2°$ which represents an acceptable value. Methods to decrease the system noise figure have been presented in chapter 3.

The concept of the up-conversion loop has been validated in the μ-wave domain, i.e. a down-conversion stage was included for the RoF measurements. In principle, it can be adapted for the mm-wave range. However, the biggest challenge the concept faces today is the scarcity of suitable dividers operable in the mm-wave range. This is a general issue in the design of mm-wave PLLs - which, in the first place, had led to the development of the translation loop.

6 Transmission Experiments

In this last part of the thesis, data transmission across MLLD-based links is investigated. We are interested in the achievable data rates in the system as it is, but also in an estimation of how much the system can be improved under optimum conditions. For this it is necessary to understand the combined effects of various system imperfections on the performance metric. The figure of merit of choice is the error vector magnitude. The two major imperfections of the MLLD-based link are additive noise as expressed in the achievable SNR (measured at the output of the system, see figure 3.4 in chapter 3) and frequency jitter induced by the MLLD (or *effective* phase noise resulting from it[1]) in an unstabilized system. We will first consider the theoretical impact of SNR and phase noise on EVM. In order to validate the theoretical findings we will continue the comparison between the MLLD-based system and the single-mode DFB system which has already been presented in chapter 5, the advantage of the latter being the possibility of varying the phase noise performance of the system. Finally, we will present a series of EVM measurements performed on the MLLD-based system for different radio standards up to a data rate of 4.234 Gbps.

6.1 Error vector magnitude

The statistical bit error as quantified by the bit error rate is the ultimate figure of merit for any type of communication system based on a digital transmission of information. It is the number of erroneous bits divided by the total number of transmitted bits in a given time interval. BER is usually estimated based on

---

[1] We have no means of separating frequency jitter and phase noise but can regard the combined effects of both as "effective" phase noise.

known signal statistics and the measurement of the "one" and "zero" rails of an eye diagram.

Minimum theoretical BER is determined by the chosen modulation format and the ratio of the energy per bit and the noise psd E_b/N_0. For example, the probability of bit error P_b for a QPSK signal is [Kam08]

$$P_{b|QPSK} = \frac{1}{2} \cdot \text{erfc}\left(\sqrt{\frac{E_b}{N_0}}\right). \tag{6.1}$$

In systems where additive noise - as quantified by SNR - is the dominant imperfection, SNR and E_b/N_0 can directly be linked. In the case of an optimum receiver filter and a white Gaussian noise channel, it is

$$SNR = \frac{E_b}{N_0} \cdot \frac{f_b}{B} \tag{6.2}$$

at a net data rate of f_b and a bandwidth B.

While this approach produces fairly reliable results in the case of simple modulation schemes, it proves increasingly complex for higher-order formats [HSS$^+$11]. Furthermore, BER is an absolute figure of merit in the sense that it tells us whether or not a given system achieves a specified performance limit. If this specified limit is not met, BER does not allow meaningful conclusions on the nature of the cause of the errors. The link designer can resort to another figure of merit, namely EVM, in order to identify the origin of transmission errors. It is usually assumed that additive noise, as quantified in the signal's SNR, is the principal limitation of a communication system, and that an increase in SNR must result in an improvement (i.e. in a reduction) of EVM. However, other imperfections that might be present during the transmission - such as gain or phase imbalances in the I- and Q-branches of modulator or demodulator, LO signal feedthrough, DC offset and phase noise - also have their signature in the constellation diagram and are reflected in EVM [Tec11].

Although transmission experiments on MLLD-based links have been presented in the past, previous works have only stated results without exposing the underlying system constraints: Huchard *et al.* have measured an EVM value of 18.5%

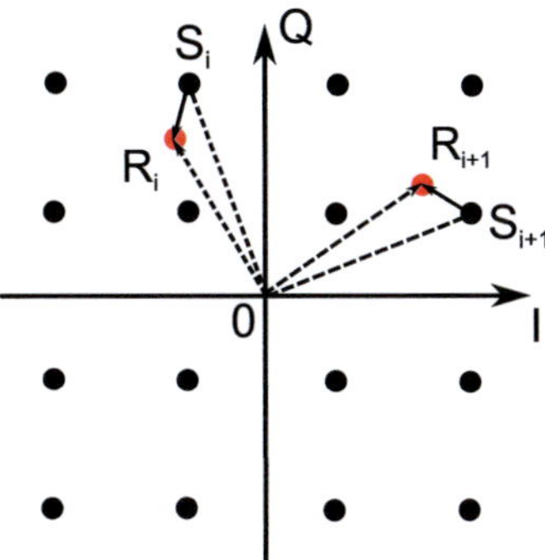

Figure 6.1: Error vector magnitude measured in a 16QAM constellation

for an SNR of 24 dB in an external modulation MLLD link providing data rates up to 3.03 Gbps [HCC$^+$08a]. The same group has reported EVM values of 10.5% for an SNR of 23 dB in a direct modulation MLLD link [HCC$^+$08b]. Despite the comparable level of SNR, a considerable difference in measured EVM values was found. This is all the more remarkable as both experiments were conducted using the same type of quantum-dash MLLD. This suggests that the mere consideration of SNR is not sufficient to obtain a reliable estimate of EVM. Based on the analyses in chapters 2 - 5, we will try to evaluate these results and link EVM performance to the system parameters that we have studied previously.

Error vector magnitude is a commonly quoted specification for both transmitter and receiver performance evaluation (e.g., ECMA 387 [Int08], IEEE 302.15.3c [IEE09]). The EVM is calculated as the root-mean-square value for a number of *I* random received symbols. It is calculated as a percentage of the average power per symbol of the constellation diagram. Figure 6.1 illustrates the error vector which for single-carrier transmission is calculated as

$$EVM_{\mathrm{rms,ap}} = \sqrt{\frac{\frac{1}{I}\sum_{n=0}^{I}|R_{\mathrm{i}} - S_{\mathrm{i}}|^2}{\frac{1}{I}\sum_{n=0}^{I}|S_{\mathrm{i}}|^2}} = \sqrt{\frac{\frac{1}{I}\sum_{n=0}^{I}|R_{\mathrm{i}} - S_{\mathrm{i}}|^2}{P_{\mathrm{avg}}}}, \qquad (6.3)$$

203

where R_i is the normalized ith symbol in the stream of measured symbols, and S_i is the ideal normalized constellation point of the ith symbol when normalized to the average symbol power P_{avg}. As a result, the recommended number of I is at least 1000 symbols [IEE09]. If the peak symbol power $|S_p|^2$ is used for normalization[2], the corresponding EVM is

$$EVM_{\mathrm{rms,pp}} = EVM_{\mathrm{rms,ap}} \cdot \left(1/\sqrt{PAV}\right) \tag{6.4}$$

with the peak-to-average symbol power ratio PAV (determined from the ideal constellation):

$$PAV = \frac{|S_p|^2}{\frac{1}{I} \sum_{n=0}^{I} |S_i|^2}. \tag{6.5}$$

For a purely phase-modulated symbol, as in PSK modulation formats, PAV is equal to 1 and both normalizations result in the same EVM value. For QAM symbols, the normalization leads to a significant difference in EVM, e.g. a factor of $PAV = 9/5 = 1.8$ for a 16QAM format, or $PAV = 17/10 = 1.7$ for 32QAM [SNW[+]12].

Table 6.1 shows reference EVM values for different data rates and different transmission types (SC, OFDM) in various existing mobile communication standards. A transmission is said to be standard-compliant if an EVM better than or at least equal to the reference value can be obtained, regardless of the properties of the channel. EVM is a standard performance metric. However, subtle differences in its calculation apply for the various configurations for which it is measured - the most relevant items being the difference between single-carrier or multi-carrier transmission, and the prevalent form of signal perturbation (additive[3], I/Q imbalance, phase noise etc.). In the following, we will briefly introduce the principal ways to calculate EVM in order to correctly interpret the measurements that will be presented later in the chapter.

---

[2] This is, e.g. the case when evaluating EVM with Agilent's VSA 89600 software.

[3] Additive white Gaussian noise (AWGN) is usually assumed, i.e. a noise which is uniform across the whole frequency spectrum and whose amplitudes follow a Gaussian distribution of a variance

Table 6.1: Reference standard EVM values

Standard document	Bit rate	Modulation	EVM (%)	Transmission
IEEE 802.11a [IEE99]	18 Mpbs	QPSK	22.39	OFDM
	24 Mpbs	16QAM	15.85	OFDM
IEEE 802.15.3c [IEE09]	Up to 1.5 Gbps	diverse	44.67	SC, OFDM
	2.1 to 2.7 Gbps	diverse	19.95	SC, OFDM
	2.8 to 5.3 Gbps	diverse	8.91	SC, OFDM
	Above 5.4 Gbps	diverse	7.08	SC, OFDM
ECMA 387 [Int08]	397 Mpbs	BPSK	33.4	in Mode A0, SC
	1588 Mbps	QPSK	23.1	in Mode A3, SC
	4234 Mbps	16QAM	11.1	in Mode A11, SC

6.1.1 Impact of SNR

If additive noise is the only or the dominant form of perturbation, EVM can be roughly estimated from the SNR ([SRRI06], [Tec11]):

$$EVM_{\mathrm{rms}} = \frac{1}{\sqrt{SNR}}. \tag{6.6}$$

A thorough analysis by Arslan and Mahmoud [AM09] has however revealed that this assumption is only valid in the case of very high values of SNR and under the condition of a data-aided receiver[4] if SNR is the principal limitation to the system. For a non-data-aided receiver used in most cases, equation (6.6) is likely to underestimate the EVM for low values of SNR or in the presence of other system imperfections. This is due to the fact that the symbol estimator will most probably assign the received symbols to their closest possible constellation point, although the deformation of the actual symbol might be much worse.

Arslan's and Mahmoud's notion of "true EVM", namely the value measured for

---

σ^2, thus representing AWGN noise power.

[4] A data-aided receiver uses known pilot signals or preambles for EVM calculation.

non-data-aided reception, corresponds better to our scenario. True EVM also depends on the SNR but in a much more complex way [AM09]:

$$EVM_{\mathrm{rms}} \approx \left[\frac{1}{SNR} - \sqrt{\frac{96}{\pi(M-1)SNR}} \sum_{i=1}^{\sqrt{M}-1} \gamma_i \cdot \exp\left(-3\beta_i^2 SNR/2(M-1)\right) \right.$$

$$\left. + \frac{12}{M-1} \sum_{i=1}^{\sqrt{M}-1} \gamma_i\beta_i \cdot \mathrm{erfc}\left(\sqrt{\frac{3\beta_i^2 SNR}{2(M-1)}} \right) \right]^{\frac{1}{2}}, \tag{6.7}$$

where $\gamma_i = 1 - i/\sqrt{M}$, $\beta_i = 2i - 1$ and $\log_2(M)$ is the number of bits coded into one symbol in an M-QAM constellation. For high values of SNR, this expression can be approximated by the simple expression in equation (6.6).

Unfortunately, equation (6.7) does not exactly reflect our scenario, as it assumes additive noise to be the only perturbation. This is not the case in our system. As we discussed in the previous chapters 2 and 5, frequency jitter and phase noise are the dominant imperfections in the unstabilized MLLD-based system. We thus need an expression which includes the contribution of both SNR and phase noise.

6.1.2 Impact of phase noise

Several studies have been published regarding different system imperfections such as fading in Rayleigh channels ([AM09]), gain and phase imbalances in the I and Q branches ([CD10], [Geo04]) and LO phase noise ([Geo04]). In the last named study, the influence of the rms phase error σ_{rms} on EVM was found to follow a simple relationship if a perfectly balanced modulator and demodulator and a Gaussian distributed probability density function can be assumed for the statistics of the phase error:

$$EVM_{\mathrm{rms}} = \sqrt{\left(\frac{1}{SNR} + 2 - 2\exp\left(\frac{\sigma_{\mathrm{rms}}^2}{2}\right) \right)}, \tag{6.8}$$

where data-aided reception was assumed. While it does not include the correction for non-data-aided reception, it is useful in the case where phase noise is the limiting factor. The phase noise present in the signal, quantified by its integrated value, thus introduces a floor to EVM, as can be observed in figure 6.2 for various values of σ_{rms} (results obtained by simulation). EVM cannot be improved by increasing SNR. For a transmission incorporating ideal oscillators, σ_{rms} is zero. The minimum EVM curve (see equation (6.6)) is reached, represented by a dashed line in the figure. The typical signature of a perturbation by phase noise is elongated arcs around the ideal symbols, as illustrated in figure 6.3 for a constellation affected by phase noise.

The disadvantage for practical EVM measurement is that we neither possess any information on the statistics of the phase error, nor can we suppose that the modulator-demodulator pair does not introduce additional imperfections.

A more pragmatic way is to determine the EVM graphically using a law of cosines[5] (refer to figure 6.4) [Ken05]. Hereby, $|S|$ is the voltage magnitude of the ideal symbol,

$$|R| = |S| + |e| \tag{6.9}$$

is the voltage magnitude of the received symbol, composed of S and an amplitude error e, and σ_{rms} is the rms phase error between them.

The amplitude error

$$|e| = N + E_{TxRx} \tag{6.10}$$

accounts for a deterministic error E_{TxRx} due to imperfections of the modulation or demodulation, as well as for random amplitude noise N. Normalized to $|S|$, the received symbol becomes with equations (6.9) and (6.10)

$$\overline{|R|} = 1 + \left(\frac{|S|}{N}\right)^{-1} + \frac{E_{TxRx}}{|S|}, \tag{6.11}$$

where $|S|/N$ denotes the SNR referred to voltages. The error vector magnitude

---

[5]In a triangle where a, b, and c are the lengths of the three sides and γ is the angle between sides a and b, the length of c can be calculated according to $c^2 = a^2 + b^2 - 2ab\cos(\gamma)$.

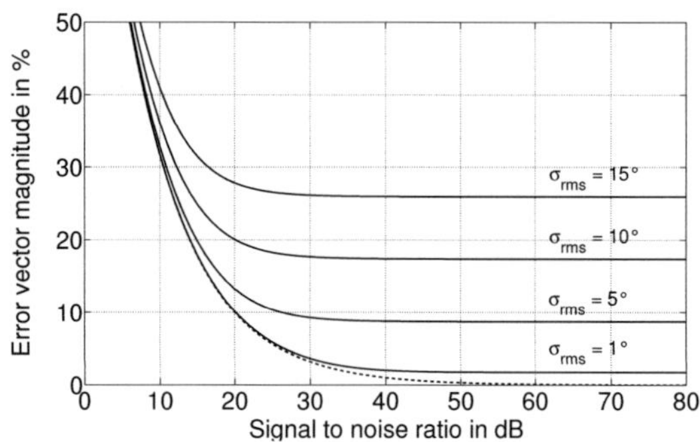

Figure 6.2: Variation of EVM with phase noise (calculated from equation (6.8))

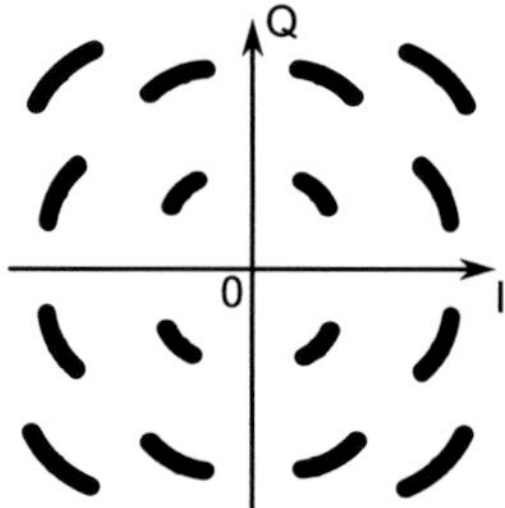

Figure 6.3: Typical signature of phase noise in a SC-16QAM signal constellation

is

$$EVM_{\mathrm{rms}} = \sqrt{1 + \overline{|R|}^2 - 2\overline{|R|}\cos(\sigma_{\mathrm{rms}})}. \qquad (6.12)$$

This approach is motivated by purely practical interest but it is nevertheless justified. We can calculate the maximum theoretical relative error between the two models ([Geo04], [Ken05]) for $E_{\mathrm{TxRx}} = 0$, referenced to [Geo04], to less than 0.45% which is a precision that we could not have otherwise attained due to the uncertainty of the EVM measurement.

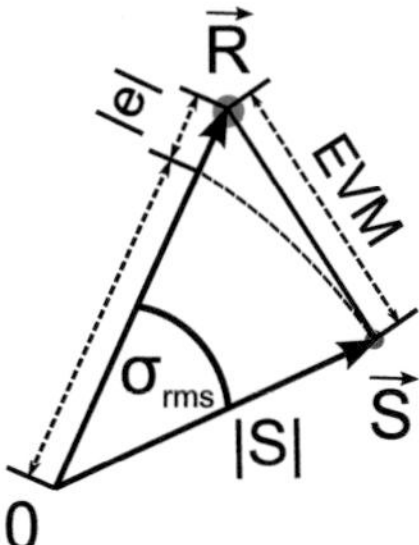

Figure 6.4: Law of cosines for EVM calculation in the presence of phase noise

6.1.3 OFDM transmission

Most modern radio standards envisage OFDM transmission. Using this technique, a wide-band channel is partitioned such that a large number of closely spaced orthogonal sub-carrier signals are used to each carry a part of the data in their own narrow-band channels. By doing so, the data rate per sub-carrier is effectively reduced. The distribution of the data across the sub-carriers has several advantages, one of which is the fact that fading due to multi-path effects or signal perturbation by narrow-band interferers only affect a small portion of the data if the remaining sub-carriers are received correctly. If error coding is implemented, it is likely that the signal can be reconstructed. In the context of OFDM transmission, EVM is typically measured over at least 20 frames and is defined as

$$EVM_{\mathrm{rms,OFDM}} = \frac{1}{N_\mathrm{p}} \sum_{i=1}^{N_\mathrm{p}} \sqrt{\frac{1}{P_\mathrm{avg} N_\mathrm{s} N_\mathrm{dsc}} \sum_{j=1}^{N_\mathrm{s}} \sum_{k=1}^{N_\mathrm{dsc}} \left| R_{\mathrm{ijk}} - S_{\mathrm{ijk}} \right|^2 }, \tag{6.13}$$

where N_p is the number of frames, N_s is the number of symbols per frame, and N_dsc is the number of data sub-carriers in the OFDM signal. In IEEE 802.15.3c, 512 sub-carriers are used, among which $N_\mathrm{dsc} = 336$ are data sub-carriers and the remaining carriers represent pilots, guards, null or DC carriers at a spacing of 4.96 MHz [IEE09]. In IEEE 802.11a (conventional WLAN), N_dsc is only 48 at a

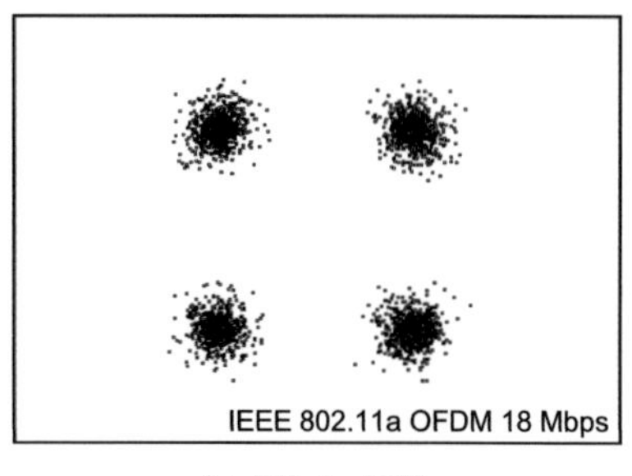

(a) Effect of ICI

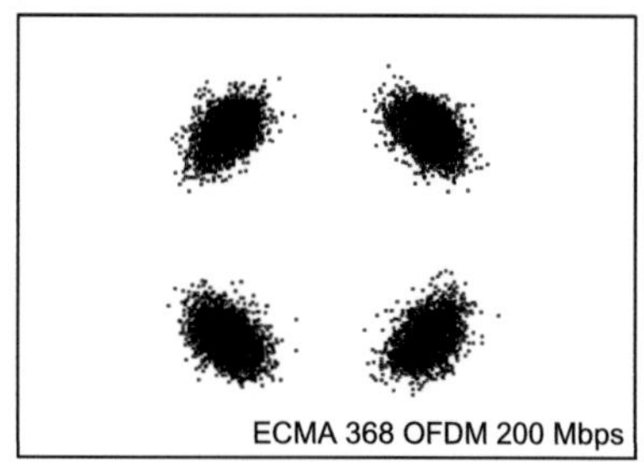

(b) Effects of ICI and CPE combined

Figure 6.5: Phase noise effects in OFDM transmission, all subcarrier superposed

carrier spacing of 250 kHz [IEE99].

Phase noise in the OFDM signal has two aspects [Sto98]: The first one is a common phase error (CPE) contribution, which implies that all carriers are rotated by the same angle simultaneously. CPE could be introduced by a local oscillator in a mixing stage. If CPE can be measured on a pilot sub-carrier bearing reference information, the data sub-carriers can be corrected accordingly. In the OFDM constellation, CPE appears as the typical phase noise signature (arcs around ideal points). The second contribution is known as inter-channel interference (ICI). ICI appears as additive noise on one channel which originates from the phase noise sidebands on other channels, affecting the first one through the demodulator function[6]. ICI appears in the OFDM constellation like any other form of additive noise as blurry spots. Figures 6.5(a) and 6.5(b) shows the effects of ICI, and the combined effects of ICI and CPE in the signal constellations, respectively.

---

[6]We can think of this phenomenon as follows: The demodulator for the ith sub-carrier has a $sinc^2$ response with zeros at the respective frequencies of the other sub-carriers. The response is however $\neq 0$ at frequency offsets between the carriers; the phase noise sidebands thus "leak through", add up and create an additive noise component on the desired sub-carrier.

210

6.2 EVM measurements on the RoF systems

If we base our EVM calculation on an rms phase error value obtained through the integration of the phase noise spectrum according to equation (5.19), we have to pay careful attention to the limits of the integration implied, especially to the upper limit. This is of course related to the fact that the EVM calculation on a vector signal analyzer will be performed on a band-limited signal. Ideally, the bandwidth around the respective carrier corresponds exactly to the bandwidth of the signal to be measured.

Before we look at the influence of the RoF system, a series of back-to-back measurements on the I/Q modulator (later feeding the E/O converter) are performed in order to get an idea of the degradation of the signal due to the RoF system.

6.2.1 Back-to-back measurements on the I/Q modulator in use

The I/Q modulator under test was the ADL5372 (Analog Devices) designed for carrier frequencies between 1500 MHz and 2500 MHz at modulation bandwidths up to 500 MHz. Using a carrier frequency of 1.8 GHz at variable phase noise levels[7] quantified by the maximum Δf of a Gaussian FM, back-to-back measurements are performed for a QPSK-modulated signal at data rates between 10 Mbps and 400 Mbps resulting in signal bandwidths between 5 MHz and 200 MHz. To this end, the I/Q modulator is connected directly to the receiver which is, in our case, a fast oscilloscope[8] which features a vector signal analyzer (VSA) package[9] for EVM measurement.

The results are shown in figure 6.6. At $\Delta f = 0$, the EVM degradation on the unperturbed carrier is clearly seen (highlighted by the ellipse). It is due to the wideband performance of the modulator itself. The larger the signal bandwidth becomes, the less marked is the relative EVM degradation with increasing phase noise. This is what we would have expected: the relative portion of the spectrum contaminated by the phase noise decreases with increasing signal bandwidth.

---

[7] As in chapter 5, the Anritsu 68377B was used.
[8] We used Agilent's Infiniium Digital Storage Oscilloscope 54855A.
[9] Agilent VSA 89600

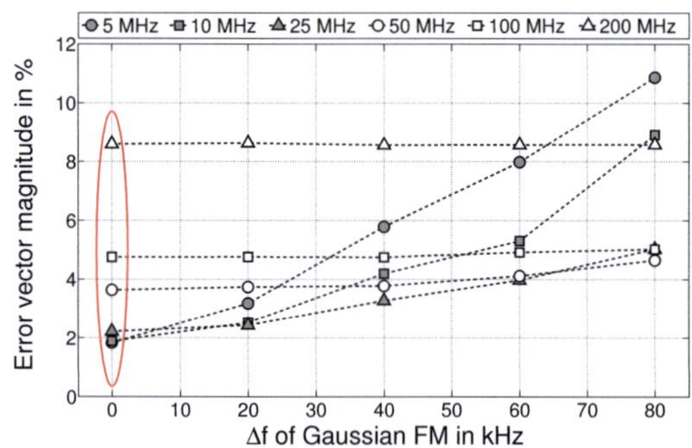

Figure 6.6: Back-to-back measurement of the EVM variation as a function of the spectral width of the modulation signal and Δf of the GN modulation

Error vector magnitude measurements can be performed on vector signal analyzers (VSAs), real-time analyzers or instruments that capture a time record of the received signal (like e.g. an oscilloscope), and internally use a fast Fourier transform (FFT) to provide frequency domain analysis. EVM measurements shown in this chapter are made using the receiver setup shown in chapter 3, figure 3.4 with the help of an oscilloscope/VSA. The oscilloscope performs a time-domain measurement. Down-conversion and signal demodulation are implemented in software. The oscilloscope emulates a radio receiver but, due to the time-based measurement, does not perfectly reproduce its behavior. In particular, signal tracking can be performed very effectively, so that relatively slow jitter in carrier frequency does not necessarily degrade the EVM result as long as the measurement time is short compared to the frequency variation of the signal. The inconvenience of such a measurement in our context is that we cannot actually evaluate the improvement achieved by stabilizing the signal frequency in terms of EVM as we do not have the possibility to disable signal tracking.

In the following, EVM measurement on both the DFB-based and the MLLD-based systems are shown for both SC and OFDM transmission. The results in this section have partially been published by the author in [BPCvD11b] and [BYP+12].

6.2.2 DFB-based system

For the DFB-based measurements, the RoF transmitter as shown in figure 5.16(b) was used. As in chapter 5, a signal generator provides the seed signal into the MZM. By frequency-modulating the generator's output signal with Gaussian noise, phase noise can deliberately be brought into the system. The frequency of the seed signal is stable. The transmission experiment is performed as **SC transmission** at an IF of f_{IF} = 1.8 GHz (see chapter 5), carrying a 50 Mbps QPSK baseband signal of pseudo-random binary data. EVM measurements on the DFB-system in figure 6.7 show a slight improvement when the PLL is active. It is most visible for very high values of GN modulation. We recall that the behavior of the MLLD-based system is reproduced for Δf = 80 kHz (see figure 5.17). The measured EVM for this value is in the order of 40% at an SNR of about 27 dB; this is also what we can expect for MLLD-based transmission.

In the following **OFDM transmission** measurements, the situation presents itself differently. It is difficult to quantify the rms phase error of the OFDM signal by a phase noise measurement as in the case for SC transmission, as we have no means to separate CPE (phase noise, as referred to in equation (6.12) and ICI contributions (phase noise that acts as additive noise). We expect EVM values to drop but cannot easily put a figure on its relative improvement. Transmission has been carried out according to the IEEE 802.11a standard (QPSK at 18 Mbps) [IEE99]. The EVM is plotted against the SNR as measured, along with the theoretical minimum curve in the absence of system imperfections (corresponding to equation (6.6)). The curves are shown in figure 6.8.

In the case of the undisturbed seed signal, EVM values of about 5% can be achieved. The SNR is limited by distortion effects of the directly modulated DFB laser (referred to as *clipping*). Below this value, the EVM closely follows the SNR curve predicted by equation (6.6), indicating that the SNR is indeed the principal limitation of the undisturbed system. For GN modulation (Δf = 80 kHz), phase noise limits the transmission even at higher SNR values and an EVM floor of about 16% is reached at about σ_{rms} = 9°. The deviation of the measured values from the theoretical curve is due to the fact that the analyzer cannot be calibrated for the reception of power values less than -45 dBm, and the analyzer so underestimates EVM.

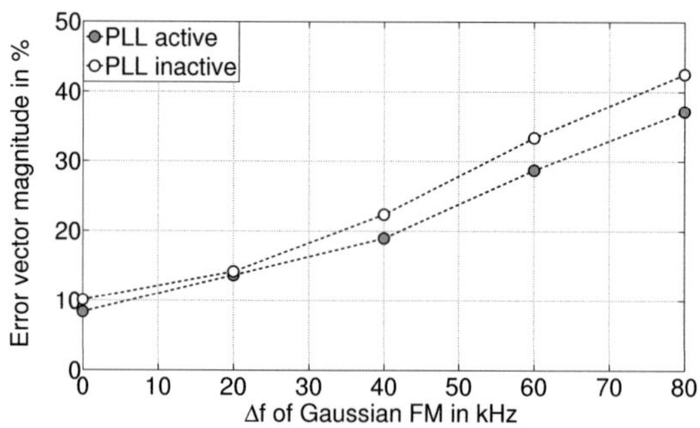

Figure 6.7: Comparison of measured values with and without PLL, DFB-based RoF system according to figure 5.16(b) (SC-QPSK 50 Mbps)

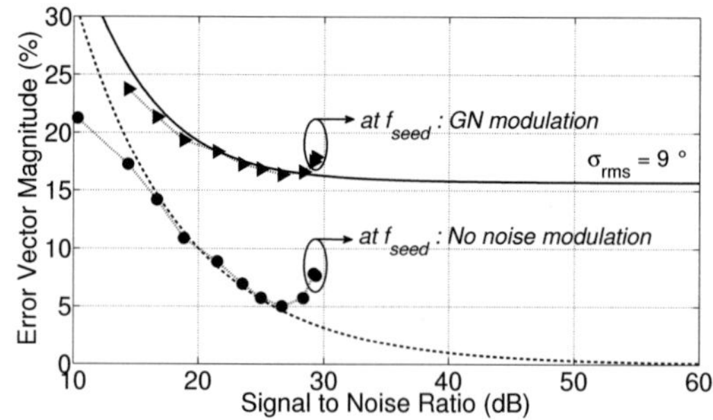

Figure 6.8: Effect of Gaussian noise modulation on an OFDM signal, DFB-based RoF system (OFDM-QPSK 18 Mbps), measured values plotted on simulated curves. Triangles: GN modulation $\Delta f = 80$ kHz. Circles: No GN modulation.

6.2.3 MLLD-based system

For the MLLD-based measurements, the external modulation RoF transmitter as shown in figure 5.16(a) was used. We recall that in this measurement, both frequency jitter and phase noise are present. As the frequency jitter is slow enough to be tracked by the scope, the former does not result in an effective degradation of EVM. This leads to the circumstance that using the PLL, we cannot improve EVM as much as would have been expected using a conventional superheterodyne test receiver: if the frequency jitter is not "seen" by the receiver, the remaining phase noise of the MLLD is "low enough" not to be improved by the realized PLL. It should however be recalled that for a realistic transmission scenario, such a result is only possible if the modulation sideband is stabilized. The stabilized lower sideband on which demodulation is performed (corresponding to the first specified band (band ID 1, f_{center} = 58.32 GHz) in both ECMA 387 and IEEE 802.15.3c standards) is shown in figure 6.9. The first transmission experiment is again a **SC transmission** at an IF of f_{IF} = 1.8 GHz at a data rate of 50 Mbps (QPSK).

The phase noise measurements in section 5.6.2 have revealed an integrated phase error of 27.53° for the MLLD-based RoF system. Applying equation (6.12), the EVM floor resulting from this value is 46.71% as shown in figure 6.10. The EVM was measured to 42.5% at an SNR of 35 dB (asterisk). The calculation slightly overestimates the measured value on the MLLD-RoF system but the order of magnitude corresponds to what we have expected from measurement on the DFB-system ($\approx$ 40%). The dashed line represents a fit of the EVM curve to our measured value, the best fit being achieved for a phase error of 25°.

In the same figure, we have plotted the measured EVM for the all-electrical measurement (circle) presented in section 5.4.3 (SNR = 31 dB, σ_{rms} = 1.50°). Here, the calculation yielding 3.85% underestimates the measured EVM value of 6.0%. At this lower integrated phase error, the EVM floor is reached only for SNR values > 45 dB. The inaccuracy of the EVM is most likely due to the imperfections of the modulator and the demodulator which are not accurately modeled by this simple approach.

We maintain that the order of magnitude is correct in both cases which, with respect to the measurement uncertainties of the setup, is highly satisfactory.

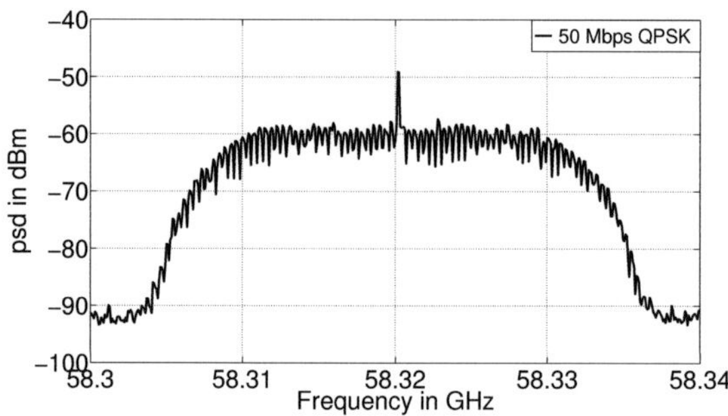

Figure 6.9: ECMA 387 / IEEE 802.15.3c Band ID 1, stabilized at 58.32 GHz center frequency (SC-QPSK 50 Mbps), RBW = 10 kHz.

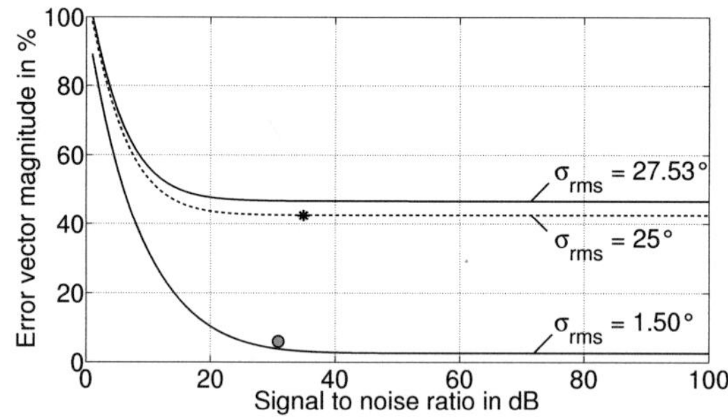

Figure 6.10: Comparison of measured values and simulated curves. Asterisk: all electrical measurement. Circle: MLLD-based system (SC-QPSK 50 Mbps).

In the **OFDM measurement**, EVM values in the order of those that have been published in ([HCC⁺08a], [HCC⁺08b]) can be achieved. Figure 6.11 shows furthermore that the MLLD-based system and the DFB-based system give the

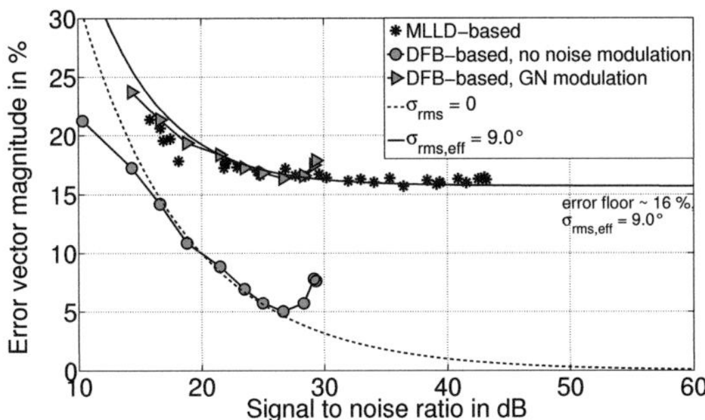

Figure 6.11: Comparison of measured values and simulated curves, both systems (OFDM-QPSK 18 Mbps)

same results in terms of EVM when the latter is GN modulated ($\Delta f = 80$ kHz), although the former reaches higher values of SNR. We thus confirm that effective phase noise is the dominant limitation to signal transmission in MLLD-based RoF systems. The EVM floor is again found at about 16%. While this value is indeed still acceptable for a QPSK WLAN transmission (see reference values in table 6.1), the transmission is likely to violate the EVM restriction provided by the standards for any higher-valued modulation format at higher data rates.

Now that we have demonstrated the general validity of our approach, the question remains as to how "good" a transmission can be made in terms of EVM if we design a better performing PLL for a more effective suppression of phase noise. The two approaches proposed in chapter 5 were:

- If all other system parameters - in particular the system noise figure responsible for the noise level outside the PLL bandwidth - remain unchanged, an integrated phase error of $19.80°$ can be achieved if the PLL is improved according to the design rules proposed in section 5.5.2. The corresponding EVM floor for sufficiently high SNR is 34.05% under the condition of SC transmission as shown above. Although a considerable improvement of a starting value of 46.71%, this value does not satisfy the specified EVM values.

- If system noise figure and hence the noise floor can effectively be reduced, **and if** all other modifications proposed in section 5.5.2 can be adhered to, a phase error of $0.21°$ can be achieved. This implies, again for SC transmission, an EVM value of 0.3%. At these levels, however, phase noise will no longer be the limiting imperfection as was made clear in the back-to-back measurements shown in section 6.2.1: the imperfect modulator and demodulator will in this case limit the transmission.

We have thus presented an estimation of the performance that can be envisaged when carefully re-designing both the MLLD-based system and the PLL.

6.2.4 Standard-compliant measurements on the MLLD-RoF system at various data rates

In the remainder of this section, we will present the results of both SC and OFDM transmission experiments as performed on the unstabilized MLLD-based RoF system. IEEE 802.11.a (WLAN) transmission[10] across the MLLD-RoF link was investigated for both direct and external modulation at 18 Mbps and 24 Mbps, modulated in OFDM-QPSK and OFDM-16QAM formats, respectively. The results are shown in figure 6.12. For a direct modulation link, the EVM shows a pronounced dependency on RF input power (and thus SNR) and a floor value of 18%. For external modulation, the comparable SNR values can be achieved at lower input power values. Therefore, the flattening of the curve due to the phase noise limitation for the external modulation link can be observed at lower input powers. The EVM reaches a floor value of around 16%. Table 6.1 shows the standard values for this type of transmission. OFDM-QPSK transmission at 18 Mbps is compliant to the standard value of 22.3% while OFDM-16QAM does not yet reach the required 15.8%.

---

[10]The WLAN signal was generated using Agilent's ADS RF System software and directly downloaded onto an Agilent ESG E4436C vector signal generator performing I/Q modulation.

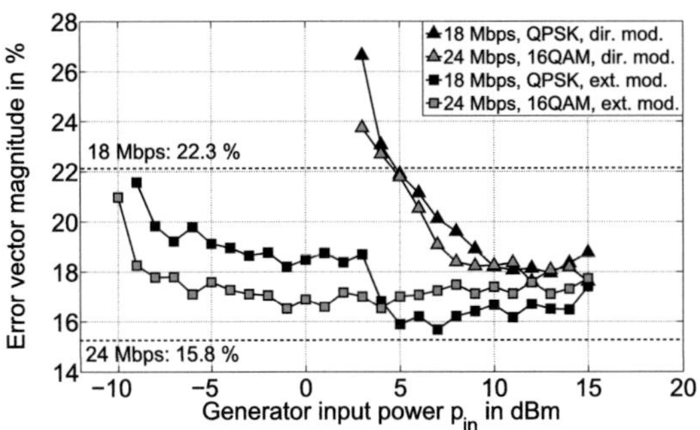

Figure 6.12: EVM as a function of modulation power (IEEE 802.11a)

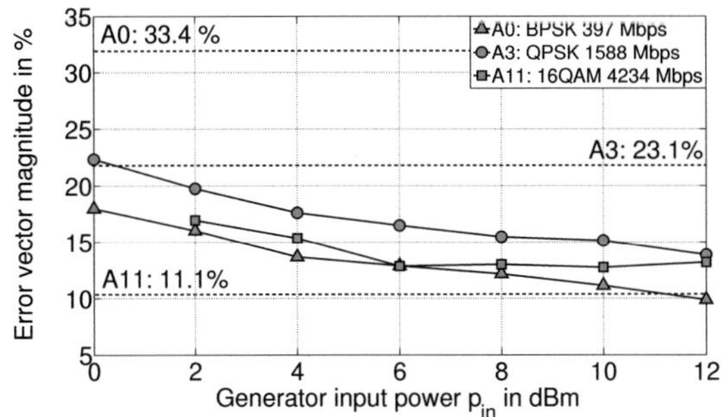

Figure 6.13: EVM as a function of modulation power (ECMA 387 / IEEE 802.15.3c)

On the external modulation link, multi gigabit[11] data transmission was performed. The test signals used for the experiment were pseudo-random binary sequences compliant with the transmission modes A0, A3 and A11 of the ECMA-387 stan-

---

[11] A Tektronix Arbitrary Waveform Generator 7211B was used to generate the digital modulation signal, and I/Q modulation was performed by an Agilent PSG E8267D.

dard, so as to evaluate three different modulation formats (SC-BPSK, SC-QPSK and SC-16QAM, respectively) at three different data rates (397 Mbps, 1588 Mbps and 4234 Mbps, respectively). Figure 6.13 sums up the results of the transmission, along with the standard EVM values for the different modes (33.4%, 23.1% and 11.1%, see also table 6.1). It should be noted that while the requirements for A11 are not yet met, standard requirements for A3 and A1 are fulfilled even at lower input power values. In all cases, the flattening of the curve due to the phase noise limitation can be observed.

6.3 Conclusions on chapter 6

This chapter was concerned with the evaluation of transmission quality in the MLLD-based RoF system. The figure of merit in this context is EVM, as it allows a qualitative analysis of the transmission error that might eventually occur. Simple EVM analysis is based on the assumption that the limiting factor is SNR, i.e. additive noise. This is no longer true in the MLLD-based system. We have presented a model for EVM which explicitly includes the contribution of phase noise. This model (see equation (6.12)) does not make any assumptions on the statistics of the phase error and is therefore easily applicable in the measurements. We have conducted transmission experiments on the MLLD-based RoF system, as well as on the DFB-based system previously presented in chapter 5. If we parametrize the latter system to exhibit a phase noise performance comparable to that of the former, we achieve comparable performance of the two systems for both SC and OFDM transmission.

While the DFB-based system is clearly limited by the achievable SNR (limited by laser distortion), the transmission experiments on the MLLD-based system confirm that this one is limited by its phase noise performance. In this respect, we refer to phase noise performance as the value of integrated rms phase error, σ_{rms}, that can be calculated from the SSB phase noise measurements shown in the previous chapter. For the high phase errors of the MLLD-based system ($\sigma_{rms} = 27.53\%$), an elevated EVM value in the order of 40 - 42% in the case of SC transmission must be expected. For higher data rates, the EVM value will drop because the relative portion of the spectrum contaminated by phase noise is reduced. The measured EVM values confirm the relationship between EVM, SNR and phase noise as established in equation (6.12).

EVM performance can be improved using OFDM techniques. EVM floor values of 16 - 18% are to be expected for the MLLD-based system. Due to the characteristics of the receiver block (time-based measurement with a scope), the relatively slow frequency jitter did not affect the measurement. This is also why the inclusion of the PLL in the MLLD-based system did not show an improving effect. In the DFB-based system, the inclusion of the PLL results in a slight improvement of EVM. On the other hand, a better performing PLL must be used for a more effective suppression of phase noise. Applying the design rules presented in the previous chapter, EVM values close to the ones measured for back-to-back transmission are feasible at sufficiently high SNR values.

Transmission experiments were conducted for two radio standards (IEEE 802.11.a, IEEE 802.15.3c). While for low data rates and spectrally inefficient modulation formats (such as BPSK), the EVM requirements can mostly be achieved, the MLLD-based system remains limited for higher-order modulation formats.

7 Conclusions and Future Prospects

The need for high data rate wireless systems and the scarcity of bandwidth in the microwave region have prompted the designers of radio communication systems to move up the spectrum and into the millimeter wave region. In Europe, up to 9 GHz bandwidth are available for unlicensed use in the 60 GHz region. The limited coverage of 60 GHz transmission is furthermore beneficial to the creation of secure small cells in indoor environments as the radio signal is naturally confined by the walls. As a direct consequence, each room needs to be equipped with its own access point. The access points are preferably passive or at least sufficiently low in complexity to be cost-effective, which is why methods of "ready-to-radiate" analog signal transmission are favored over the transmission of digital baseband signals to the access point. However, the electrical delivery of an analog signal in the mm-wave range is highly lossy and therefore rather expensive and inflexible. The quasi-lossless optical fiber is the natural choice for the guided transmission of such signals. Optical signal generation and modulation can be achieved by the remote heterodyning of at least two optical wavelengths separated by the desired mm-wave frequency. Mode-locked laser diodes represent a class of multi-mode lasers which deliver coherent modes for optical heterodyning, and which can replace more complex multi-laser or laser-modulator architectures commonly employed in remote heterodyning. InAs/InGaAsP quantum-dash structures have been identified as particularly interesting for the integration of such diodes into miniature packages. Theoretical analyses of the physics of quantum-dash mode-locked laser diodes have predicted advantageous component features, such as high intrinsic modulation bandwidth and narrow beat note linewidths. Transmission experiments have demonstrated the principal feasibility of MLLD-enabled multi gigabit transmission across short-haul links. On the other hand, the properties of mode-locked laser diodes had not yet been studied systematically in the context of mm-wave fiber-optic links. In particular, there had not yet been an analysis linking component features to system-relevant parameters such as link gain or noise figure which would in turn provide a specification for the design of components dedicated for use as mm-wave generators.

From the findings of the previous chapters, we can draw a set of conclusions that allow us to put into perspective the future use of mode-locked laser diodes for mm-wave RoF systems. In this work, we have presented a comprehensive component characterization of the mode-locked laser diodes with respect to their qualification as optoelectronic oscillators when combined with an appropriate high speed detector. In particular, we have pointed out the principal aspects that should receive the full attention of both device and system designers. The power delivered by the MLLD-based system in both carrier and sideband components is strongly dependent on the spectral width of the laser, or its number of modes, as well as on the available optical power. It is equally important to optimize chip-to-fiber coupling by ensuring a very low level of optical retro-reflections into the laser cavity. These aspects, and those related, were treated in chapters 3 and 4. The precision with which the desired mm-wave carrier frequency can be adjusted during device manufacturing is not yet sufficient to qualify for high-precision oscillators. Rather than the actual carrier frequency, the crucial issue is however the high frequency jitter of up to 600 kHz uncertainty at best in the generated mm-wave carrier. The level of bias current as well as the chip-to-fiber coupling play a vital role here. Due to the jitter of the carrier frequency, the net phase noise performance of the mode-locked laser diode cannot reliably be measured at close offset frequencies. Frequency and phase noise related aspects were studied in chapters 5 and 6.

Furthermore, a system-level study on link architectures was presented, in particular a comparison of direct and external MLLD modulation links, where the relevant system parameters were derived and validated by experiment.
The most relevant result is again related to the spectral width of the laser: the intrinsic link gain scales in inverse proportion to the number of modes. For both direct and external modulation links, the link gain can be increased effectively through a limitation of the mode-locked laser spectrum. At a given optical power, the net improvement from limiting the spectrum from about 30 - 40 modes to

only 20 modes lies in the range of 26 dB. For direct modulation, a trade-off could be identified between the stability of the generated beat note (high bias current) and the efficiency of the E/O conversion (low bias current in the linear region). Bandwidth-independent improvement of link gain can be achieved through an improvement of laser slope efficiency, i.e. more available optical power at a given bias point. Through impedance matching, a bandwidth-dependent improvement can be obtained. In the external modulation link, the conversion efficiency is limited by the modulator's design parameters and can in principle be increased independently from the bias of the laser. In practice, it is limited by the input power the mode-locked laser diode can provide.

For both direct and external modulation links, the low power delivered by the system requires the use of an electrical amplifier. The system noise figure in the range of 65 dB to 70 dB approaches the passive attenuation limit. Due to the high negative link gain, the system noise figure is hardly affected by the electrical amplifier. The system noise figure and the signal-to-noise ratio resulting therefrom can only be improved through a considerable increase in link gain. All efforts to improve intrinsic link gain should therefore be undertaken. Among these, the limitation of the spectrum to a small number of modes is the most promising, as it affects the power budgets of direct and external modulation links alike. With regard to the linearity and distortion performance of the links, it was found that the direct modulation link is linear up to a maximum modulation power into the laser starting from which the mode-locking is perturbed. Intermodulation distortion could not be measured. In the external modulation link, the dynamic range is limited by the transfer function of the chosen modulator. In conclusion, the performance of state-of-the-art mode-locked laser diodes has yet to be increased according to the design criteria that have been derived in this study. External link modulation architectures can be employed to mitigate the effects of the limited E/O conversion of the diode.

Another study was dedicated to the propagation effects in mode-locked laser based communication systems, especially concerning the effects of chromatic dispersion, and the periodical power loss resulting therefrom. We assert that the absolute transmission distance in the MLLD-system is not limited by dispersion effects; allowed link lengths are limited to discrete values depending on the mm-wave carrier frequency generated by the diode and the chirp performance. For

60 GHz transmission, the periodicity of the power maxima is about 2 kilometers. The chirp-induced shift of the first power maximum can typically be found in 60 - 70 meters distance. Implications for the link designer have been stated.

The frequency stability of the MLLD-generated carrier wave is of great concern, as it translates onto the sideband in the up-conversion process, regardless of the stabilization measures that might eventually have been undertaken in the IF range. The problem of frequency jitter was approached by the development of a phase-locked loop architecture enabling the stabilization of the modulation sideband in the presence of high carrier phase noise or carrier frequency instability.

A system demonstrator was implemented and tested. The technique can always be employed when a DSB electrical spectrum is generated at some point in the transmitter. The concept of sideband stabilization could be validated both on an all-electrical system, where the double sideband spectrum was created from mixing with an LO signal provided by a signal generator, and on the laser-based radio-over-fiber system. Two optical sub-systems were investigated: a DFB-laser based system with stable carrier center frequency, but additional phase noise modulation; and the MLLD-based system exhibiting both frequency jitter and a certain level of phase noise. The center frequency of a sideband in such a RoF system can thus be synthesized to a precise value. The proposed technique can be applied to all DSB systems where the LO is optically generated.

In conclusion, the problems of frequency instability and that of the mediocre precision of frequency selection can be considered to be solved. Through simulation, the potential improvement of the phase-locked loop demonstrator itself, and of its performance when employed with the MLLD-based system, was estimated, the most important aspect being again system noise figure (or intrinsic link gain).

Finally, the transmission experiments carried out for data rates up to 4.234 Gbps have shown that standard-compliant SC and OFDM signal transmission is feasible in the MLLD-system, yet limited by the aspects that we have already identified. In particular, the relationship between the rms phase error and the error vector magnitude was clarified; for an rms error of about $9°$, we must expect an EVM limitation in the order of about 15% (OFDM transmission), independent from the signal-to-noise ratio. Design criteria for a high-performance standard-compliant system were derived.

Future research

We have pointed out techniques to improve the performance of MLLD-based RoF systems. Their key advantage over other architectures remain the small size of the MLLD and its potential for integration. On the other hand, much of this is lost when reverting to system architectures employing off-chip $LiNbO_3$ modulators, filters, amplifiers or other external components. Much effort is therefore devoted to exploring the possibilities of integrating these functions with the laser diode. We have seen that the separation of mm-wave signal generation for up-conversion and E/O conversion is beneficial to the link performance in an external modulation link. It is an obvious idea to try and combine both functions on one chip in order to avoid bulk devices while at the same time profiting from external modulation. In a first study at III-V Lab, Joint Laboratory of Alcatel Lucent Bell Labs, Thales Research & Technology [vD12], the FP laser architecture was replaced by a DBR laser and integrated with an on-chip EAM. First characterization results have shown promising results, but the EAM is potentially more sensitive to retro-reflections from the fiber. In an effort to mitigate both the effects of power loss due to a large number of higher harmonics and GVD dispersion, attempts have been made to limit the spectrum from 15 nm to about 8 nm (i.e. from >40 to about 20 modes). It was observed that neither the frequency jitter nor the phase noise characteristics of the device were altered.

A last modification to the laser chips concerns the coupling performance of the diodes where on-chip tapers are being studied. The possible power gain from tapering the output is believed to be as high as 10 dB, yielding output power of at least 10 to 15 dBm on the fiber.

Further development toward a better performing structure can also be imagined at system level. In the 60 GHz band, where sufficient bandwidth is *a priori* available, it might be envisioned to move away from complex digital modulation and instead opt for simple on-off-keying schemes. The receiver architecture can then be implemented as simple Schottky detector as used in electrical AM systems. While the maximum achievable data rates will be limited with respect to complex (I, Q) modulation schemes, an advantage of this architecture is its robustness in relation to frequency or phase noise.

A Photo-Detection

The photo-detector (PD) is principally a semiconductor diode structure which ideally should exhibit high bandwidth, low current flow in the absence of an optical signal ("dark current"), weak internal noise, and high sensitivity. At the same time, the structure should lend itself to the integration with other optical or electrical components. Typically, heterostructures based on indium gallium arsenide (InGaAs) on indium phosphite (InP) substrate are used for operation in the telecom window around 1.55 µm. A common architecture is the p-i-n diode (p-doped - intrinsic - n-doped), operated in the photoconductive mode, i.e. under reverse bias conditions. Ideally, the incident photons are absorbed in the intrinsic zone only, whereat the (short) length of the intrinsic zone determines the effective (high) cut-off frequency of the diode [Fre09].

All measurements presented in this work were obtained with a p-i-n diode[1] with a bandwidth of 70 GHz.

A circuit model of the diode is shown in figure A.1, where V_b is the bias voltage, v_d and i_d are the voltage across and the current through the diode, G_P is the parasitic parallel conductance of the diode, R_S is its series bulk resistance, and $R_a = 50\ \Omega$ is the output resistance of the diode matched to the standard line impedance of a coaxial cable loading the diode, $R_0 = 50\ \Omega$. Note that the diode is biased in reverse. The capacitors $C_{1,2}$ and the resistor R_1 correspond to the external circuitry of the diode, according to the manufacturer's datasheet.

If the diode is operated in photoconductive mode, and V_b is far below the breakdown voltage, and $R_a \gg R_S$, the differential diode resistance including $1/G_P$ is far bigger than the output resistance - this is generally the case. If, moreover, the photocurrent exceeds by far the dark current of the structure, then the current i_a will equal the generated photocurrent i. With a dark current in the order of a few nA and a photocurrent of (typically) several mA, this assumption is usually valid.

---

[1] U2T photonic's 70 GHz photo-detector XPDV3120R

The photocurrent i is furthermore proportional to the received optical power p_{opt} with a factor ρ, the so-called *responsivity* of the photo-detector. The received optical power, in turn, is the product of the effective detector surface and the optical intensity, and thus, proportional to the squared magnitude of the electrical field of the incident wave $\left|e_{\mathrm{opt}}\right|^2$.

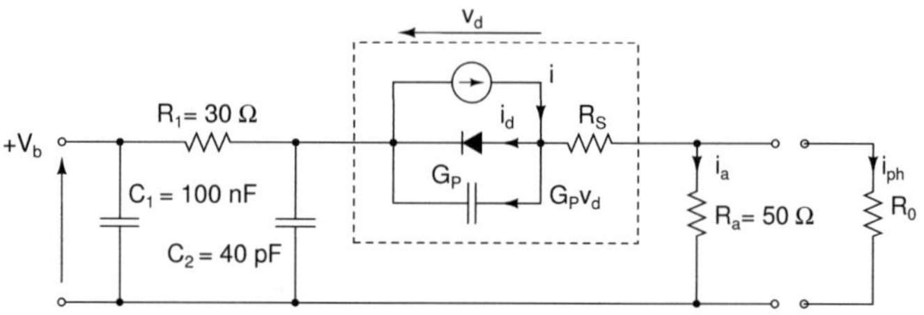

Figure A.1: Circuit model of photo-detector

Under these conditions, the current through the matched load R_0 is

$$i_{\mathrm{ph}} = \frac{R_0}{R_0 + R_{\mathrm{a}}} \cdot \rho\, p_{\mathrm{opt}} = \frac{1}{2}\rho\, p_{\mathrm{opt}}, \tag{A.1}$$

yielding an electrical power[2] of

$$P_{\mathrm{el}} = R_0 \left\langle i_{\mathrm{ph}} \right\rangle^2. \tag{A.2}$$

The responsivity ρ in amperes per watts (A/W) determines the performance of the opto-electrical conversion process, or the efficiency of photo-detection. It is determined by the external quantum efficiency η_{D}, and the ratio of the electron charge to the energy of the incident photon at λ_0 and c_0:

$$\rho = \frac{\eta_{\mathrm{D}} q \lambda_0}{h c_0} \tag{A.3}$$

In an ideal detection process ($\eta_{\mathrm{D}} = 1$), where every incident photon is converted

---

[2]The $\langle\rangle$ takes the rms value of the quantity.

Table A.1: Features of the photo-detector used in this work [pA10]

Parameter	Condition	Typical Value
Operating wavelength range	-	1480 .. 1620 nm
Average optical input power	-	-20 .. 13 dBm
Reverse voltage $-V_b$	-	2.8 V
Responsivity ρ	optimum polarization	0.6 A/W
Polarization dependent loss	-	0.3 dB
3 dB cut-off frequency	measurement with heterodyning system	75 GHz
Output resistance R_a	-	50 Ω
Dark current	-	200 nA

into an electron, the maximum responsivity at 1.55 µm is $\rho = 1.24$ A/W. Real devices typically reach $\rho = 0.6 - 0.9$ A/W ([MHMW91],[WSPH91]).

B The Mach-Zehnder modulator

In the domain of fiber-optic communication systems, the Mach-Zehnder modulator is the most common type of external electro-optical converter[1]. Thanks to the linear electro-optic effect in non-centrosymmetric materials, also referred to as the *Pockels* effect, a beam of incoming light can be modulated by an electrical signal. Its principle of operation and the underlying physics shall briefly be explained here. The concept of modulation efficiency will be presented in the context of a Mach-Zehnder modulator, and the most common modulator architectures will be discussed.

The Pockels Effect

In 1906, Friedrich Pockels noted that a variation of the refractive index n can be induced in an optical medium when a constant or varying electrical field is applied, whereat the variation is proportional to the field [Pas08]. The phenomenon since carries his name, and the Pockels cell has been put to use in numerous applications, most prominently, in the making of optical phase or intensity modulators. These devices are based on uniaxial crystal structures, such as lithium niobate ($LiNbO_3$) or lithium tantalate ($LiTaO_3$), i.e., the crystal is anisotropic exhibiting an ordinary index n_o for light polarized in the two directions x and y, and an extraordinary index n_e for polarization in a third direction z.

For $LiNbO_3$ devices, the Pockels effect is best exploited in the extraordinary axis, yielding the largest coefficient of the non-linear material tensor, in the order of $r_{33} = 30\,\mathrm{pm/V}$ [Ari04]. If an electrical field E_z is applied in z-axis, resulting from a voltage V on the electrodes of length L and separated by d, the index variation induces a phase modulation of the optical wave passing through the device

---

[1]Other types of modulators included electro-absorption modulators, or acusto-optic modulators.

$$\Delta\varphi = \frac{\pi L}{\lambda} n_{\text{eff}}^3 r_{33} E_z = \frac{\pi L}{\lambda} n_{eff}^3 r_{33} \frac{V}{d},$$ (B.1)

where n_{eff} is the effective refractive index associated with the optical wave of vacuum wavelength λ. The Pockels effect has only a small dependence on wavelength or temperature and will respond to frequencies higher than 100 GHz [CIABP06]. An important modulator parameter from a link design standpoint is the voltage required to induce 180 degrees, or π radians, of phase shift between the two optical waves at the output combiner; this voltage is referred to as the half-wave voltage V_π.

$$V_\pi = \frac{\lambda}{n_{\text{eff}}^3 r_{33}} \frac{d}{L}.$$ (B.2)

V_π depends therefore on the physical dimensions of the device. It is the defining quantity for the modulator's transfer curve. The phase variation $\Delta\varphi(t)$ can be stated in terms of the applied voltage $v(t)$ and V_π,

$$\Delta\varphi(t) = \frac{\pi v(t)}{V_\pi}.$$ (B.3)

Intensity modulation

When the Pockels cell is inserted into one of the two arms of a Mach-Zehnder interferometer, see figure B.1, the phase variation translates into an intensity modulation of the optical wave. This is achieved by splitting the optical field at a first Y-junction, letting the two resulting waves propagate through the two arms, whereas one of the two experiences an additional phase shift $\Delta\varphi(t)$, and recombining them at a second Y-junction. By controlling the phase variation via the linear electro-optic effect, the interference and thus, the output intensity, is varied. The architecture shown in figure B.1 is referred to as single-drive configuration. A dual-drive configuration comprises a Pockels cell in both arms, and the modulator is driven by two voltage signals ("push-pull"). As a single-electrode MZM was used in this work, we will in the following concentrate on the single-drive configuration.

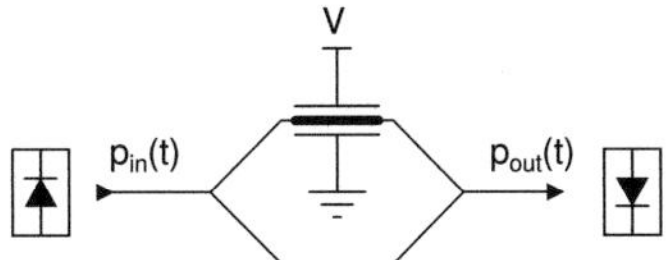

Figure B.1: Pockels cell in Mach-Zehnder interferometer

For a perfectly balanced and lossless modulator, the wave is split and recombined as

$$e_{\text{out}}(t) = E_{\text{in}} \left(\exp\left(j\frac{2\pi}{\lambda}t \right) + \exp\left(j\frac{2\pi}{\lambda}t + \Delta\varphi(t) \right) \right). \tag{B.4}$$

Transfer function and modulation efficiency

The transfer function H_{m} of the modulator relates the power $p_{\text{m,out}}$ at the output to the power p_{in} at the input of the modulator; it is a function of the applied voltage $v(t)$:

$$H_{\text{m}} = \frac{p_{\text{m,out}}}{p_{\text{in}}} = \frac{T_{\text{i}}}{2}\left(1 + \cos\left(\frac{\pi}{2}\frac{v(t)}{V_{\pi}} \right) \right), \tag{B.5}$$

where T_{i} is the fiber-to-fiber transmission of the modulator. Intensity modulation is achieved by operating the modulator at a bias point V_{B} and applying a small-signal voltage modulation $v_{\text{m}}(t)$,

$$v(t) = V_{\text{B}} + v_{\text{m}}(t). \tag{B.6}$$

For the static regime, $v(t) = V_{\text{B}}$ is constant. H_{m} then is fundamentally non-linear. However, if the bias point V_{B} is chosen at the quadrature point, i.e. the inflection point of the $\cos^2$- curve, the transfer function can be linearized, and the device operates as a linear modulator. It is then $V_{\text{B}} = (2k+1)V_{\pi}/2$, and we chose $k = 0$. Thus, for small-signal RF modulation the following approximation applies:

$$p_{m,opt} = \frac{T_i p_{in}}{2} \left(-\frac{\pi v_{m,rms}}{V_\pi} \right),$$ (B.7)

where $v_{m,rms}$ is the rms value of the modulating signal $v_m(t)$ [CI04]. The frequency response of the MZM is considered flat and the electrode is matched to the source impedance. In terms of power, we therefore consider the MZM an ohmic impedance of R_m. For a matched MZM, the input impedance R_m equals the source impedance R_g and, with the voltage divider rule, the power delivered by the generator is

$$p_{g,a} = \frac{v_g^2}{4R_g} = \frac{v_{m,rms}^2}{R_m}.$$ (B.8)

For the modulation efficiency we find

$$\frac{p_{m,opt}^2}{p_{g,a}} = \left(\frac{T_i p_{in} \pi}{2V_\pi} \right)^2 \cdot R_m,$$ (B.9)

and the equivalent MZM fiber-couple slope efficiency at this particular bias point

$$s_{mzm} \left(V_B = \frac{V_\pi}{2} \right) = \frac{T_i p_{in} \pi R_m}{2V_\pi}.$$ (B.10)

At least two more points of interest can be identified. At the point of *minimum transmission*, the modulator suppresses the linear modulation and produces the second harmonic of the modulation.

Modulator architectures

In designing an MZM, a choice must be made regarding the orientation of the waveguides and electrodes to the crystals, commonly referred to as "cut". The cut affects the characteristic parameters of the modulator, such as V_π and the level of modulator chirp α_m. The strongest component of the applied electric field must be aligned with the extraordinary axis of the crystal which exhibits the strongest electro-optic coefficient (r_{33}). For an x-cut configuration, this requires that the waveguide be placed between the electrodes. For a z-cut device, the waveguide must lie beneath the electrodes. In order to minimize the attenuation of the optical mode due to metal absorption, z-cut devices require a buffer layer. The buffer layer is not necessarily needed in an x-cut device; however, it is often included for velocity matching of the RF and the optical wavex for very high-speed operation [WKYY$^+$00]. The most common electrode configurations for MZM are shown in figure B.2. X-cut topologies result in chirp-free operation due to the inherent push-pull symmetry of the applied fields in the electrode gaps. Using z-cut topologies, the chirp value can be controlled by adjusting the drive levels in a dual-drive configuration. Balancing the two drive levels results in chirp-free operation while a purely single-drive z-cut devices exhibit chirp values of about -0.7 [WKYY$^+$00].

Frequency response and matching

The maximum modulation frequency of the linear electro-optic effect in lithium niobate is in the hundreds of GHz. In contrast to laser diodes, where the maximum modulation frequency is dependent on both material and device parameters, the upper limit on the Mach-Zehnder modulation frequency is not set by the material. It is rather the design and the architecture which limit the modulation performance of an MZM. Also unlike the laser diode, the modulator impedance - and hence the frequency response - is independent of modulator bias [CI04]. As we have seen in equation (B.10), the modulation efficiency of the MZM-based link depends on the value of V_π. In order to achieve low half-wave voltages, the modulator electrodes need to be made longer - which in turns increases the capacitance of the (lumped element) electrodes and so reduces the achievable bandwidth. Higher bandwidth can then be obtained by employing travelling wave electrodes. The electrical signal propagates along the same direction as the optical wave. The bandwidth is in this case limited by the mismatch between the electrical and optical propagation constants. Another limiting factor is the electrical loss of the electrode due to the attenuation of high frequencies [WKYY$^+$00].

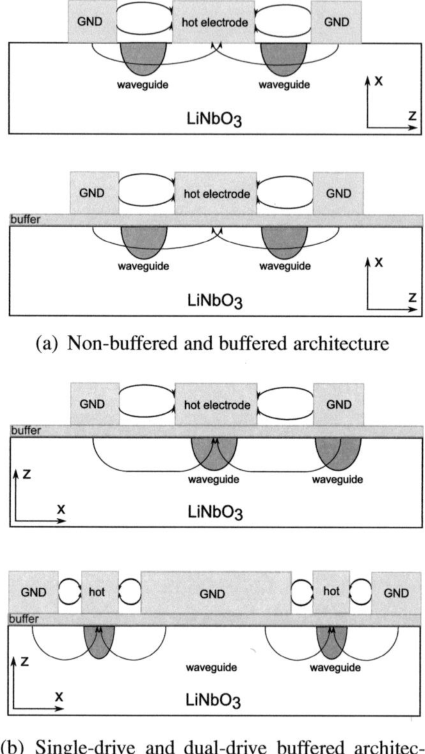

(a) Non-buffered and buffered architecture

(b) Single-drive and dual-drive buffered architectures

Figure B.2: Modulator architectures: (a) x-cut, (b) z-cut. From [WKYY$^+$00].

Bibliography

[ABB+05]   E.I. Ackerman, G. Betts, W. Burns, J. Prince, M. Regan, H. Roussell, and C.H. Cox III. Low noise figure, wide bandwidth analog optical link. In *International Topical Meeting on Microwave Photonics (MWP), Seoul, Korea*, pages 325 – 328, Oct 2005.

[ACIB+98]   E.I. Ackerman, C.H. Cox III, G. Betts, H. Roussell, F. O'Donnell, and K. Ray. Input impedance conditions for minimizing the noise figure of an analog optical link. *IEEE Transactions on Microwave Theory and Techniques*, 46(12):2025 –2031, Dec 1998.

[ACIR98]   E.I. Ackerman, C.H. Cox III, and N.A. Riza. *Selected Papers on Analog Fiber-Optic Links*. Number v. MS 149 in SPIE Milestone Series. SPIE Press, 1998.

[Agr97]   G.P. Agrawal. *Fiber-Optic Communication Systems*. Wiley Series in Microwave and Optical Engineering. John Wiley & Sons, 2nd edition, 1997.

[Agr05]   G. Agrawal. *Lightwave Technology - Telecommunication Systems*. John Wiley & Sons, 1st edition, 2005.

[All10]   Wireless Gigabit Alliance. WiGig White Paper: Defining the future of multi-gigabit wireless communications. `http://www.wirelesshd.org/pdfs/WirelessHD-Specification-Overview-v1.1May2010.pdf`, Jul 2010. Online; accessed 18-Sep-2012.

[ALN+96]   Z. Ahmed, H.F. Liu, D. Novak, Y. Ogawa, M.D. Pelusi, and D.Y. Kim. Locking characteristics of a passively mode-locked monolithic DBR laser stabilized by optical injection. *IEEE Photonics Technology Letters*, 8(1):37 –39, Jan 1996.

[AM97]   V.F. Andrievski and S.A. Malyshev. High speed InGaAs photode-
            tector modules for fibre optic communications. In *Workshop on
            High Performance Electron Devices for Microwave and Optoelec-
            tronic Applications, 1997. EDMO 1997*, pages 335 –339, Nov
            1997.

[AM09]   H. Arslan and H.A. Mahmoud. Error vector magnitude to SNR
            conversion for nondata-aided receivers. *IEEE Transactions on
            Wireless Communications*, 8(5):2694 –2704, May 2009.

[AMP00]   E.A. Avrutin, J.H. Marsh, and E.L. Portnoi. Monolithic and multi-
            gigahertz mode-locked semiconductor lasers: Constructions, ex-
            periments, models and applications. *IEE Proceedings - Optoelec-
            tronics*, 147(4):251–278, 2000.

[AMTO03]   S. Arahira, N. Mineo, K. Tachibana, and Y. Ogawa. 40 GHz
            hybrid modelocked laser diode module operated at ultra-low RF
            power with impedance-matching circuit. *IET Electronics Letters*,
            39(3):287 – 289, Feb 2003.

[ANWL97]   Z. Ahmed, D. Novak, R.B. Waterhouse, and H.F. Liu. 37 GHz
            fiber-wireless system for distribution of broad-band signals. *IEEE
            Transactions on Microwave Theory and Techniques*, 45(8):1431
            –1435, Aug 1997.

[Ari04]   L. Arizmendi. Photonic applications of lithium niobate crystals.
            physica status solidi (a), 201(2):253–283, 2004.

[ARLZ08]   S.I. Alekseev, A.A. Radzievsky, M.K. Logani, and M.C. Ziskin.
            Millimeter wave dosimetry of human skin. *Wiley Bioelectromag-
            netics Journal*, 29(1):65–70, 2008.

[AS82]   Y. Arakawa and H. Sakaki. Multidimensional quantum well laser
            and temperature dependence of its threshold current. *Applied
            Physics Letters*, 40(11):939–941, 1982.

[Ban06]   D. Banerjee. *PLL Performance, Simulation, and Design*. National Semiconductors, 4th edition, 2006.

[BB84]   C. Bulmer and W. Burns. Linear interferometric modulators in Ti:LiNbO$_3$. *IEEE Journal of Lightwave Technology*, 2(4):512 – 521, Aug 1984.

[Bes07]   R.E. Best. *Phase-locked loops - Design, Simulation, and Applications*. McGraw Hill, 6th edition, 2007.

[BGL98]   D. Bimberg, M. Grundmann, and N. Ledentsov. *Quantum Dot Heterostructures*. John Wiley & Sons, 1st edition, 1998.

[BGP$^+$93]   L.A. Buckman, J.B. Georges, J. Park, D. Vassilovski, J.M. Kahn, and K.Y. Lau. Stabilization of millimeter-wave frequencies from passively mode-locked semiconductor lasers using an optoelectronic phase-locked loop. *IEEE Photonics Technology Letters*, 5(10):1137 –1140, Oct 1993.

[BGvH$^+$96] R.P. Braun, G. Grosskopf, C.H. von Helmolt, K. Kruger, U. Kruger, D. Rohde, and F. Schmidt. Optical microwave generation and transmission experiments in the 12 and 60 GHz-region for wireless communications. *Microwave Symposium Digest, 1996., IEEE MTT-S International*, 2:499–501 vol.2, Jun 1996.

[Bod45]   H.W. Bode. *Network Analysis and Feedback Amplifier Design*. Van Nostrand, New York, 1945.

[BPC$^+$11a] F. Brendel, J. Poette, B. Cabon, T. Zwick, F. Lelarge, and F. van Dijk. Analog link performance of mode-locked laser diodes in the 60 GHz range. In *International Topical Meeting on Microwave Photonics (MWP)/ Asia-Pacific Microwave Photonics Conference (APMP), Singapore*, pages 153 –156, Oct 2011.

[BPC$^+$11b] F. Brendel, J. Poette, B. Cabon, T. Zwick, F. van Dijk, F. Lelarge, and A. Accard. Chromatic dispersion in 60 GHz radio-over-fiber networks based on mode-locked lasers. *IEEE Journal of Lightwave Technology*, 29(24):3810 –3816, Dec 2011.

[BPCvD11a] F. Brendel, J. Poette, B. Cabon, and F. van Dijk. Génération de porteuses rf par diodes laser a verrouillage de modes pour les réseaux
personnels hybrides a 60 GHz. In *Journées Nationales Microon-
des (JNM), Rennes, France*, May 2011.

[BPCvD11b] F. Brendel, J. Poette, B. Cabon, and F. van Dijk. Low-cost analog
fiber optic links for in-house distribution of millimeter-wave signals. *Cambridge/EuMA International Journal of Microwave and
Wireless Technologies*, 3:231–236, 2011.

[BYP$^+$12] F. Brendel, J. Yi, J. Poette, B. Cabon, and T. Zwick. Properties
of millimeter-wave signal generation and modulation using mode-
locked Q-dash lasers for gigabit RF-over-fiber links. In *7th German Microwave Conference (GeMiC), Ilmenau, Germany*, pages
1 –4, Mar 2012.

[BZP$^+$12] F. Brendel, T. Zwick, J. Poette, B. Cabon, and F. van Dijk. PLL-
based stabilization of optically generated millimeter-wave carrier
signals in radio-over-fiber systems. To be submitted to the *IEEE
Journal of Lightwave Technology*, 2012.

[BZPC12] F. Brendel, T. Zwick, J. Poette, and B. Cabon. Modulation sideband stabilization in the presence of carrier frequency jitter and
phase noise. Submitted to the *IEEE Transactions of Microwave
Theory and Techniques*, 2012.

[CA05] Australian Communications and Media Authority (ACMA). Radiocommunications (low interference potential devices) class li-
cense variation. `http://www.acma.gov.au/WEB/STANDARD/pc=
PC_297`, Aug 2005. Online; accessed 18-Sep-2012.

[Cam98] E. Camargo. *Design of FET frequency multipliers and harmonic
oscillators*. Artech House Microwave Library. Artech House,
Boston, 13th edition, 1998.

[CB06]   A. Chorti and M. Brookes. A spectral model for RF oscillators with power-law phase noise. *IEEE Transactions on Circuits and Systems I: Regular Papers*, 53(9):1989–1999, Sep 2006.

[CCC$^+$08] B. Charbonnier, P. Chanclou, J. L. Corral, G. H. Duan, C. Gonzalez, M. Huchard, D. Jäger, F. Lelarge, J. Marti, L. Naglic, L. Pavlovic, V. Polo, R. Sambaraju, A. Steffan, A. Stöhr, M. Thual, A. Umbach, F. Van Dijk, M. Vidmar, and M. Weiss. Photonics for broadband radio communications at 60 GHz in access and home networks. In *International Topical Meeting on Microwave Photonics (MWP)/ Asia-Pacific Microwave Photonics Conference (APMP), Gold Coast, Australia*, 2008.

[CD10]   Z.Q. Chen and F.F. Dai. Effects of LO phase and amplitude imbalances and phase noise on M -QAM transceiver performance. *IEEE Transactions on Industrial Electronics*, 57(5):1505–1517, May 2010.

[Cha11]   B. Charlet. *Etude et réalisation de sources laser impulsionnelles en optique intégrée sur verre : application á la génération de supercontinuum.* PhD thesis, Université de Grenoble, 2011.

[CI04]   C.H. Cox III. *Analog Optical Links - Theory and Practice.* Cambridge University Press, 1st edition, 2004.

[CIABP06]   C.H. Cox III, E.I. Ackerman, G.E. Betts, and J.L. Prince. Limits on the performance of RF-over-fiber links and their impact on device design. *IEEE Transactions on Microwave Theory and Techniques*, 54(2):906–920, Feb 2006.

[CIAHB97]   C.H. Cox III, E.I. Ackerman, R. Helkey, and G.E. Betts. Direct-detection analog optical links. *IEEE Transactions on Microwave Theory and Techniques*, 45(8):1375–1383, Aug 1997.

[CIRRH98]   C.H. Cox III, H.V. Roussell, R.J. Ram, and R.J. Helkey. Broadband, directly modulated analog fiber link with positive intrinsic gain and reduced noise figure. In *International Topical Meeting on Microwave Photonics (MWP), Princeton, New Jersey, USA*, pages 157 –160, Oct 1998.

[CLB$^+$10]   N. Chimot, F. Lelarge, R. Brenot, A. Accard, J.G. Provost, and H. Debregeas. 1550 nm directly modulated lasers for 10 Gb/s SMF transmission up to 65 km at 45° with chirp optimized InAs/InP quantum dashes. In *36th European Conference and Exhibition on Optical Communication (ECOC), Torino, Italy*, pages 1 –3, Sep 2010.

[Com04]   Federal Communications Commission. Code of Federal Regulation title 47 Telecommunication, Chapter 1, part 15.255. `http://ecfr.gpoaccess.gov/cgi/t/text/text-idx?c=ecfr&tpl=/ecfrbrowse/Title47/47cfr15_main_02.tpl`, Oct 2004. Online; accessed 19-Sep-2012.

[Cor12]   Corning. Product information: Single mode optical fiber. `http://www.corning.com/WorkArea/showcontent.aspx?id=41261` , 2012. Online; accessed 19-Sep-2012.

[CTPW09]   G. Carpintero, M.G. Thompson, R.V. Penty, and I.H. White. Low noise performance of passively mode-locked 10 GHz quantum-dot laser diode. *IEEE Photonics Technology Letters*, 21(6):389 –391, Mar 2009.

[CUYBC95]   T.R. Chen, J. Ungar, X.L. Yeh, and N. Bar-Chaim. Very large bandwidth strained MQW DFB laser at 1.3 µm. *IEEE Photonics Technology Letters*, 7(5):458 –460, May 1995.

[dB32]   H. de Bellescize. La réception synchrone. *L'Onde Électrique*, 11:230–240, Jun 1932.

[DB90]   T.E. Darcie and G.E. Bodeep. Lightwave subcarrier CATV transmission systems. *IEEE Transactions on Microwave Theory and Techniques*, 38(5):524 –533, May 1990.

[Dem02]   A. Demir. Phase noise and timing jitter in oscillators with colored-noise sources. *IEEE Transactions on Circuits and Systems I: Fundamental Theory and Applications*, 49(12):1782 – 1791, Dec 2002.

[Des94]   E. Desurvire. *Erbium Doped Fiber Amplifiers: Principles and Applications*. John Wiley & Sons, 1st edition, 1994.

[DMR00]   A. Demir, A. Mehrotra, and J. Roychowdhury. Phase noise in oscillators: a unifying theory and numerical methods for characterization. *IEEE Transactions on Circuits and Systems I: Fundamental Theory and Applications*, 47(5):655 –674, May 2000.

[Ega99]   W.F. Egan. *Frequency Synthesis by Phase Lock*. 2nd edition. John Wiley & Sons, 1999.

[Ega00]   W.F. Egan. *Phase Lock Basics*. John Wiley & Sons, 2nd edition, 2000.

[Ega03]   W.F. Egan. *Practical RF System Design*. John Wiley & Sons, 1st edition, 2003.

[ERC11]   ERC. ERC recommendation 70-03 (tromso 1997 and subsequent amendments) relating to the use of short range devices (SRD). `http://www.erodocdb.dk/Docs/doc98/official/pdf/REC7003E.pdf`, Aug 2011. Online; accessed 19-Sep-2012.

[EWAD88]   A.F. Elrefaie, R.E. Wagner, D.A. Atlas, and D.G. Daut. Chromatic dispersion limitations in coherent lightwave transmission systems. *IEEE Journal of Lightwave Technology*, 6(5):704–709, May 1988.

[FAB$^+$10]   G. Fiol, D. Arsenijevic, D. Bimberg, A.G. Vladimirov, M. Wolfrum, E.A. Viktorov, and P. Mandel. Hybrid mode-locking in a 40 GHz monolithic quantum dot laser. *Applied Physics Letters*, 96(1):011104, 2010.

[Fan50]   R.M. Fano. Theoretical limiations on the broadband matching of arbitrary impedances. *Journal of the Franklin Institute*, 249(1):57–83, 1950.

[Fis02]   U. Fischer. *Optoelectronic Packaging*. VDE Verlag, 1st edition, 2002.

[FLK97]   J.C. Fan, C.L. Lu, and L.G. Kazovsky. Dynamic range requirements for microcellular personal communication systems using analog fiber-optic links. *IEEE Transactions on Microwave Theory and Techniques*, 45(8):1390–1397, Aug 1997.

[Fre09]   W. Freude. *Optische Empfänger und Fehlerwahrscheinlichkeit, Lecture Notes*. Institut für Photonik und Quantenelektronik, Karlsruher Institut für Technologie, 2009.

[Fri44]   H.T. Friis. Noise figures of radio receivers. *Proceedings of the IRE*, 32(7):419 – 422, Jul 1944.

[Gar79]   F.M. Gardner. *Phaselock Techniques*. John Wiley & Sons, 2nd edition, 1979.

[Gar80]   F.M. Gardner. Charge-pump phase-lock loops. *IEEE Transactions on Communications*, 28(11):1849 – 1858, Nov 1980.

[Geo04]   A. Georgiadis. Gain, phase imbalance, and phase noise effects on error vector magnitude. *IEEE Transactions on Vehicular Technology*, 53(2):443 – 449, Mar 2004.

[GG90]   A. Ghiasi and A. Gopinath. Novel wide-bandwidth matching technique for laser diodes. *IEEE Transactions on Microwave Theory and Techniques*, 38(5):673 –675, May 1990.

[GK93]   C.L. Goldsmith and B. Kanack. Broad-band reactive matching of high-speed directly modulated laser diodes. *IEEE Microwave and Guided Wave Letters*, 3(9):336 –338, Sep 1993.

[GM07]   M. Gioannini and I. Montrosset. Numerical analysis of the frequency chirp in quantum-dot semiconductor lasers. *IEEE Journal of Quantum Electronics*, 43(10):941 –949, Oct 2007.

[GMA09]   N. Ghazisaidi, M. Maier, and C. Assi. Fiber-wireless (FiWi) access networks: A survey. *IEEE Communications Magazine*, 47(2):160 –167, Feb 2009.

[GMM⁺06]   C. Gosset, K. Merghem, G. Moreau, A. Martinez, G. Aubin, J.L. Oudar, A. Ramdane, and F. Lelarge. Phase-amplitude characterization of a high-repetition-rate quantum dash passively mode-locked laser. *OSA Optics Letters*, 31(12):1848–1850, 2006.

[GNN96]   U. Gliese, S. Norskov, and T.N. Nielsen. Chromatic dispersion in fiber-optic microwave and millimeter-wave links. *IEEE Transactions on Microwave Theory and Techniques*, 44(10):1716–1724, Oct 1996.

[GSLO99]   R.A. Griffin, H.M. Salgado, P.M. Lane, and J.J. O'Reilly. System capacity for millimeter-wave radio-over-fiber distribution employing an optically supported PLL. *IEEE Journal of Lightwave Technology*, 17(12):2480, Dec 1999.

[GT00]   A. Ghatak and K. Thyagarajan. *Introduction to Fiber Optics*. Cambridge University Press, 1st edition, 2000.

[HAB⁺59]   H. A. Haus, W.R Atkinson, G.M. Branch, W.B Davenport, W.H. Fonger, W.A. Harris, S.W. Harrison, W.W. McLeod, E.K. Stodola, and D.E. Talpey. IRE standards on methods of measuring noise in linear twoports, 1959. *Proceedings of the IRE*, 48(1):60 –68, Jan 1959.

[Hau00]   H.A. Haus. Mode-locking of lasers. *IEEE Journal of Selected Topics in Quantum Electronics*, 6(6):1173–1185, Nov/Dec 2000.

[HBMM04]   P.K. Hanumolu, M. Brownlee, K. Mayaram, and Un-Ku Moon. Analysis of charge-pump phase-locked loops. *IEEE Transactions on Circuits and Systems I: Regular Papers*, 51(9):1665 – 1674, Sep 2004.

[HBTH]   F. Hakimi, M.G. Bawendi, R. Tumminelli, and J.R. Haavisto. US Patent No. 5260957: Quantum Dot Laser. Issued on Nov 9, 1993.

[HCC$^+$08a]   M. Huchard, P. Chanclou, B. Charbonnier, F. van Dijk, G.-H. Duan, C. Gonzalez, F. Lelarge, M. Thual, M. Weiss, and A. Stohr. 60 GHz radio signal up-conversion and transport using a directly modulated mode-locked laser. In *International Topical Meeting on Microwave Photonics (MWP)/ Asia-Pacific Microwave Photonics Conference (APMP), Gold Coast, Australia*, pages 333–335, Oct 2008.

[HCC$^+$08b]   M. Huchard, B. Charbonnier, P. Chanclou, F. Van Dijk, F. Lelarge, G.H. Duan, C. Gonzalez, and M. Thual. Millimeter-wave photonic up-conversion based on a 55 GHz quantum dashed mode-locked laser. In *34th European Conference and Exhibition on Optical Communication (ECOC), Brussels, Belgium*, Sep 2008.

[Hen82]   C. Henry. Theory of the linewidth of semiconductor lasers. *IEEE Journal of Quantum Electronics*, 18(2):259 – 264, Feb 1982.

[Her98]   F. Herzel. An analytical model for the power spectral density of a voltage-controlled oscillator and its analogy to the laser linewidth theory. *IEEE Transactions on Circuits and Systems I: Fundamental Theory and Applications*, 45(9):904 –908, Sep 1998.

[HHN03]   A. Hirata, M. Harada, and T. Nagatsuma. 120 GHz wireless link using photonic techniques for generation, modulation, and emission of millimeter-wave signals. *IEEE Journal of Lightwave Technology*, 21(10):2145 – 2153, Oct 2003.

[HKM77]   S. Hata, K. Kajiyama, and Y. Mizushima. Performance of p-i-n photodiode compared with avalance photodiode in the longer-wavelength region of 1 to 2 μm. *IET Electronics Letters*, 13(22):668 –669, 27 1977.

[HMW+11]   W. Hofmann, M. Müller, P. Wolf, A. Mutig, T. Gründl, G. Böhm, D. Bimberg, and M.C. Amann. 40 Gbit/s modulation of 1550 nm VCSEL. *IET Electronics Letters*, 47(4):270–271, 17 2011.

[HOR+09]   T. Habruseva, S. O'Donoghue, N. Rebrova, F. Kéfélian, S.P. Hegarty, and G. Huyet. Optical linewidth of a passively mode-locked semiconductor laser. *OSA Optics Letters*, 34(21):3307–3309, 2009.

[HSS+11]   D. Hillerkuss, R. Schmogrow, T. Schellinger, M. Jordan, M. Winter, G. Huber, T. Vallaitis, R Bonk, P. Kleinow, F. Frey, M. Roeger, S. Koenig, A. Ludwig, A. Marculescu, J. Li, M. Hoh, M. Dreschmann, J. Meyer, S. Ben Ezra, N. Narkiss, B. Nebendahl, F. Parmigiani, P. Petropoulos, B. Resan, A. Oehler, K. Weingarten, T. Ellermeyer, J. Lutz, M. Moeller, M. Huebner, J. Becker, W. Koos, W. Freude, and J. Leuthold. 26 Tbit s^{-1} line-rate superchannel transmission utilizing all-optical fast Fourier transform processing. *Nature Photonics*, 5:364371, 2011.

[IEE93]   IEEE Standard American National Standard Canadian Standard Graphic Symbols for Electrical and Electronics Diagrams (Including Reference Designation Letters), 1993.

[IEE99]   IEEE. Supplement to IEEE standard for information technology - telecommunications and information exchange between systems - local and metropolitan area networks - specific requirements. part 11: Wireless LAN medium access control (MAC) and physical layer (PHY) specifications: High-speed physical layer in the 5 GHz band. *IEEE Std 802.11a-1999*, page i, 1999.

[IEE09]   IEEE. IEEE standard for information technology - telecommunications and information exchange between systems - local and metropolitan area networks - specific requirements. part 15.3: Wireless medium access control (MAC) and physical layer (PHY) specifications for high rate wireless personal area networks

(WPANs) amendment 2: Millimeter-wave-based alternative physical layer extension. *IEEE Std 802.15.3c-2009 (Amendment to IEEE Std 802.15.3-2003)*, pages c1 −187, 12 2009.

[IEE12]   IEEE. IEEE 802.11.ad: Very high throughput in 60 GHz. `http://www.ieee802.org/11/Reports/tgad_update.htm`, expected for Dec 2012. Online; accessed 18-Sep-2012.

[Ins]   European Telecommunications Standards Institute. ETSI TR 102 555: Electromagnetic compatibility and radio spectrum matters (ERM): System reference document: Technical characteristics of multiple gigabit wireless systems in the 60 GHz range.

[Int08]   ECMA International. ECMA-387: High rate 60 GHz PHY, MAC and HDMI PAL. `http://www.ecma-international.org/publications/standards/Ecma-387.htm`, Dec 2008. Online; accessed 18-Sep-2012.

[Kam08]   K.D. Kammeyer. *Nachrichtenübertragung*. Vieweg und Teubner, 4th edition, 2008.

[KD87]   B.H. Kolner and D.W. Dolfi. Intermodulation distortion and compression in an integrated electrooptic modulator. *OSA Applied Optics*, 26(17):3676–3680, Sep 1987.

[Ken05]   P.B. Kenington. *RF and Baseband Techniques for Software Defined Radio*. Mobile Communication Series. Artech House, 1st edition, 2005.

[KK06]   K. Kammeyer and K. Kroschel. *Digital Signalverarbeitung: Filterung und Spektralanalyse mit MATLAB-Übungen*. B.G. Teubner Fachverlag, Wiesbaden, 6th edition, 2006.

[KMS⁺04]   F.X. Kärtner, U. Morgner, T. Schibli, R. Ell, H.A. Haus, J.G. Fujimoto, and E.P. Ippen. *Few-cycle laser pulse generation and its applications*, chapter Few-cycle pulses directly from a laser, pages 73–135. Topics in Applied Physics. Springer Berlin, 2004.

[Koo06]   T. Koonen. Fiber to the home/fiber to the premises: What, where, and when? *Proceedings of the IEEE*, 94(5):911–934, May 2006.

[KOT+08]   F. Kéfélian, S. O'Donoghue, M.T. Todaro, J.G. McInerney, and G. Huyet. RF linewidth in monolithic passively mode-locked semiconductor laser. *IEEE Photonics Technology Letters*, 20(16):1405–1407, Aug 2008.

[KSS01]   K. Kojucharow, M. Sauer, and C. Schaffer. Millimeter-wave signal properties resulting from electrooptical upconversion. *IEEE Transactions on Microwave Theory and Techniques*, 49(10):1977–1985, Oct 2001.

[KT96]   Y. Katagiri and A. Takada. Synchronised pulse-train generation from passively mode-locked semiconductor lasers by a phase-locked loop using optical modulation sidebands. *IET Electronics Letters*, 32(20):1892–1894, Sep 1996.

[KVT+11]   J. Kreissl, V. Vercesi, U. Troppenz, T. Gaertner, W. Wenisch, and M. Schell. Up to 40 Gb/s directly modulated laser operating at low driving current: Buried-heterostructure passive feedback laser (BH-PFL). *IEEE Photonics Technology Letters*, PP(99):1, 2011.

[KYJ+08]   A.M.J. Koonen, H. Yang, H.-D. Jung, Y. Zheng, J. Yang, H.P.A. van den Boom, and E. Tangdiongga. Recent research progress in hybrid fiber-optic in-building networks. *Joint conference of the Opto-Electronics and Communications Conference, 2008 and the 2008 Australian Conference on Optical Fibre Technology. OECC/ACOFT 2008.*, pages 1–2, Jul 2008.

[Lac10]   K. Lacanette. Application note 779: A basic introduction to filters: Active, passive, and switched-capacitor. `http://www.ti.com/lit/an/snoa224a/snoa224a.pdf`, 2010. Online; accessed 19-Sep-2012.

[LDR+07]   F. Lelarge, B. Dagens, J. Renaudier, R. Brenot, A; Accard, F. Van Dijk, D. Make, O. Le Gouezigou, J.-G. Provost, F. Point,

J. Landreau, O. Drisse, E. Derouin, B. Rousseau, F. Pommereau, and G-H. Duan. Recent advances on InAs/InP quantum dash based semiconductor lasers and optical amplifiers operating at 1.55 µm. *IEEE Journal of Selected Topics in Quantum Electronics*, 13(1):111 – 124, Jan/Feb 2007.

[Lee66] D.B. Leeson. A simple model of feedback oscillator noise spectrum. *Proceedings of the IEEE*, 54(2):329 – 330, Feb 1966.

[LGL$^+$11] C.Y Lin, F. Grillot, Y. Li, R. Raghunathan, and L.F. Lester. Microwave characterization and stabilization of timing jitter in a quantum-dot passively mode-locked laser via external optical feedback. *IEEE Journal of Selected Topics in Quantum Electronics*, 17(5):1311 –1317, Sep/Oct 2011.

[LKON06] M.G. Larrode, A.M.J. Koonen, J.J.V. Olmos, and A. Ng'Oma. Bidirectional radio-over-fiber link employing optical frequency multiplication. *IEEE Photonics Technology Letters*, 18(1):241 – 243, Jan 2006.

[LTC$^+$09] F. Lecoche, E. Tanguy, B. Charbonnier, H.W. Li, F. van Dijk, A. Enard, F. Blache, M. Goix, and F. Mallecot. Transmission quality measurement of two types of 60 GHz millimeter-wave generation and distribution systems. *IEEE Journal of Lightwave Technology*, 27(23):5469 –5474, Dec 2009.

[LZS$^+$08] E.K Lau, X. Zhao, H.K. Sung, D. Parekh, C. Chang-Hasnain, and M.C. Wu. Strong optical injection-locked semiconductor lasers demonstrating $\leq$ 100 GHz resonance frequencies and 80 GHz intrinsic bandwidths. *OSA Optics Express*, 16(9):6609–6618, Apr 2008.

[MB06] Z. Mi and P. Bhattacharya. DC and dynamic characteristics of p-doped and tunnel injection 1.65 µm InAs quantum-dash lasers grown on InP (001). *IEEE Journal of Quantum Electronics*, 42(12):1224 –1232, Nov 2006.

[MC07]   S.A. Malyshev and A.L. Chizh. Pin photodiodes for frequency mixing in radio-over-fiber systems. *IEEE Journal of Lightwave Technology*, 25(11):3236–3243, Nov 2007.

[MHMW91]   M. Makiuchi, H. Hamaguchi, T. Mikawa, and O. Wada. Easily manufactured high-speed back-illuminated GaInAs/InP p-i-n photodiode. *IEEE Photonics Technology Letters*, 3(6):530 –531, Jun 1991.

[MIMH08]   N. Mohamed, S.M. Idrus, A.B. Mohammad, and H. Harun. Millimeter-wave carrier generation system for radio over fiber. In *International Symposium on High Capacity Optical Networks and Enabling Technologies (HONET)*, pages 111–115, Nov 2008.

[MRA$^+$09]   K. Merghem, R. Rosales, S. Azouigui, A. Akrout, A. Martinez, F. Lelarge, G. Aubin, G.H. Duan, and A. Ramdane. Low noise performance of passively mode locked quantum-dash-based lasers under external optical feedback. *Applied Physics Letters*, 95:131111–1 –3, 2009.

[MT05]   Canada Spectrum Management and Telecommunications. Radio Standard Specification-210, issue 6, Low-power licensed-exempt radio communication devices (all frequency bands): Category 1 equipment. `http://www.ic.gc.ca/eic/site/smt-gst.nsf/eng/h_sf06129.html`, Sep 2005. Online; accessed 19-Sep-2012.

[MTHM79]   T. Miya, Y. Terunuma, T. Hosaka, and T. Miyashita. Ultimate low-loss single-mode fibre at 1.55 µm. *IET Electronics Letters*, 15(4):106 –108, 15 1979.

[MYJ64]   G.L. Matthaei, L. Young, and E.M.T. Jones. *Microwave Filters, Impedance Matching Networks, and Coupling Structures.* McGraw-Hill New York, 1st edition, 1964.

[NCLG09]   G.H. Nguyen, B. Cabon, and Y. Le Guennec. Generation of 60-GHz MB-OFDM signal-over-fiber by up-conversion using cascaded external modulators. *IEEE Journal of Lightwave Technology*, 27(11):1496–1502, 2009.

[NCP$^+$09]   H.G. Nguyen, B. Cabon, J. Poette, Z. Yu, and P.Y. Fonjallaz. All optical up-conversion of WLAN signal in 60 GHz range with sideband suppression. In *Radio and Wireless Symposium, 2009. RWS '09. IEEE*, pages 590 –593, Jan 2009.

[NDCL09]   G.H. Nguyen, V. Dobremez, B. Cabon, and Y. LeGuennec. Optical techniques for up-conversion of MB-OFDM signals in 60 GHz band using fiber Bragg grating. In *IEEE International Conference on Communications (ICC)*, 2009.

[NFP$^+$10]   A. Ng'oma, D. Fortusini, D. Parekh, W. Yang, M. Sauer, S. Benjamin, W. Hofmann, M.C. Amann, and C.J. Chang-Hasnain. Performance of a Multi-Gb/s 60 GHz radio over fiber system employing a directly modulated optically injection-locked VCSEL. *IEEE Journal of Lightwave Technology*, 28(16):2436 –2444, Aug 2010.

[NII09]   T. Nagatsuma, H. Ito, and T. Ishibashi. High-power RF photodiodes and their applications. *Wiley Laser and Photonics Reviews*, 3:123–137, 2009.

[NMC$^+$93]   Y. Nakano, M.L. Majewski, L.A. Coldren, H.L. Cao, K. Tada, and H. Hosomatsu. Intrinsic modulation response of a gain-coupled MQW DFB laser with an absorptive grating. In *Integrated Photonics Research*, page ITuG6. Optical Society of America, 1993.

[NSGF91]   D.K. Negus, L. Spinelli, N. Goldblatt, and G. Feugnet. Sub-100 femtosecond pulse generation by Kerr lens mode-locking in Ti:Al$_2$O$_3$. In *Advanced Solid State Lasers*, page SPL7. Optical Society of America, 1991.

[Nyq28]   H. Nyquist. Thermal agitation of electric charge in conductors. *Physical Review*, 32:110–113, Jul 1928.

[Off]   European Radiocommunications Office. Website. `http://www.ero.dk`. Online; accessed 19-Sep-2012.

[OJ82]   H. Olesen and G. Jacobsen. A theoretical and experimental analysis of modulated laser fields and power spectra. *IEEE Journal of Quantum Electronics*, 18(12):2069 – 2080, Dec 1982.

[Olv89]   A.D. Olver. Millimetre wave systems-past, present and future. *Radar and Signal Processing, IEE Proceedings F*, 136(1):35 – 52, Feb 1989.

[oPA09]   European Conference of Postal and Telecommunications Administrations. ECC recommendation (09)01: Use of the 57 - 64 GHz frequency band for point-to-point fixed wireless systems. `http://www.erodocdb.dk/docs/doc98/official/pdf/Rec0901.pdf`, Jan 2009. Online; accessed 19-Sep-2012.

[oPMHA00]   Ministry of Public Management and Telecommunications Home Affairs, Posts. Japan specified low power radio station (12) 59-66 GHz band, regulation for the enforcement of the radio law 6-4-2., 2000.

[pA10]   U2T photonics AG. Datasheet 70 GHz photodetector XPDV3120R. `http://www.u2t.com/index.php/en/products/all-products/item/xpdv3120r`, 2010. Online; accessed 19-Sep-2012.

[Pas08]   R. Paschotta. *Encyclopedia of Laser Physics and Technology*. John Wiley & Sons, 2008.

[PGMA98]   J.K. Piotrowski, B.A. Galwas, S.A. Malysev, and V.F. Andrievski. Investigation of InGaAs p-i-n photodiode for optical-microwave mixing process. In *12th International Conference on Microwaves and Radar (MIKON)*, pages 171 –175 vol.1, May 1998.

[PKS11]   M. Park, K.C. Kim, and J.I. Song. Generation and transmission of a quasi-optical single sideband signal for radio-over-fiber systems. *IEEE Photonics Technology Letters*, 23(6):383 –385, Mar 2011.

[Poz05]   D.M. Pozar. *Microwave Engineering*. John Wiley & Sons, 3rd edition, 2005.

[RB10]   E. Rubiola and R. Boudot. Phase noise in RF and microwave amplifiers. In *Frequency Control Symposium (FCS), 2010 IEEE International*, pages 109 –111, Jun 2010.

[RBD$^+$05] J. Renaudier, R. Brenot, B. Dagens, F. Lelarge, B. Rousseau, F. Poingt, O. Legouezigou, F. Pommereau, A. Accard, P. Gallion, and G.H. Duan. 45 GHz self-pulsation with narrow linewidth in quantum dot fabry-perot semiconductor lasers at 1.5 μm. *IET Electronics Letters*, 41(18):1007–1008, Sep 2005.

[RCS07]   E.U. Rafailov, M.A. Cataluna, and W. Sibbett. Mode-locked quantum-dot lasers. *Nature Photonics*, 1:395–401, Jul 2007.

[Rub09]   E. Rubiola. *Phase Noise and Frequency Stability in Oscillators*. The Cambridge RF and Microwave Engineering Series. Cambridge University Press, 2009.

[SAH$^+$09] H.-J. Song, K. Ajito, A. Hirata, A. Wakatsuki, T. Furuta, N. Kukutsu, and T. Nagatsuma. Multi-gigabit wireless data transmission at over 200 GHz. In *34th International Conference on Infrared, Millimeter, and Terahertz Waves (IRMMW-THz)*, pages 1 –2, Sep 2009.

[SAPB10]   H. Shams, P.M. Anandarajah, P. Perry, and L.P. Barry. Optical generation of modulated millimeter waves based on a gain-switched laser. *IEEE Transactions on Microwave Theory and Techniques*, 58(11):3372 –3380, Nov 2010.

[SBC$^+$10] A. Stoehr, S. Babiel, P.J. Cannard, B. Charbonnier, F. van Dijk, S. Fedderwitz, D. Moodie, L. Pavlovic, L. Ponnampalam, C.C. Renaud, D. Rogers, V. Rymanov, A. Seeds, A.G. Steffan, A. Umbach, and M. Weiss. Millimeter-wave photonic components for broadband wireless systems. *IEEE Transactions on Microwave Theory and Techniques*, 58(11):3071 –3082, Nov 2010.

[Sch18]   W. Schottky. Über spontane Stromschwankungen in verschiedenen Elektrizitätsleitern. *Annalen der Physik*, 362(23):541–567, 1918.

[SGR$^+$]   A. Shen, Ch. Gosset, J. Renaudier, G. H. Duan, J-L. Oudar, F. Lelarge, F. Pommereau, F. Poingt, L. le Gouezigou, and 0. le Gouezigou. Ultra-narrow mode-beating spectral line-width of a passively mode-locked quantum dot fabry-perot laser diode. In *32th European Conference and Exhibition on Optical Communication (ECOC), Cannes, France*.

[Sha49]   C.E. Shannon. Communication in the presence of noise. *Proceedings of the IRE*, 37(1):10 – 21, Jan 1949.

[SHR73]   J.H. Shoaf, D. Halford, and A.S. Risley. *Frequency Stability Specification and Measurement: High Frequency and Microwave Signals*. US Department of Commerce / National Bureau of Standards, 1973.

[Smu02]   P. Smulders. Exploiting the 60 GHz band for local wireless multimedia access: Prospects and future directions. *IEEE Communications Magazine*, 40(1):140 –147, Jan 2002.

[SN79]   M. Saruwatari and K. Nawata. Semiconductor laser to single-mode fiber coupler. *OSA Applied Optics*, 18(11):1847–1856, Jun 1979.

[SNA97]   G.H. Smith, D. Novak, and Z. Ahmed. Overcoming chromatic-dispersion effects in fiber-wireless systems incorporating external modulators. *IEEE Transactions on Microwave Theory and Techniques*, 45(8):1410 –1415, Aug 1997.

[SNW$^+$12] R. Schmogrow, B. Nebendahl, M. Winter, A. Josten, D. Hillerkuss, S. Koenig, J. Meyer, M. Dreschmann, M. Huebner, C. Koos, J. Becker, W. Freude, and J. Leuthold. Error vector magnitude as a performance measure for advanced modulation formats. *IEEE Photonics Technology Letters*, 24(1):61 –63, Jan 2012.

[SPLG$^+$12] T. Shao, F. Paresys, Y. Le Guennec, G. Maury, N. Corrao, and B. Cabon. Photonic generation and radio transmission of ECMA 387 signal at 60 GHz using WDM demultiplexer. *Wiley Microwave and Optical Technology Letters*, 54(2):275–277, 2012.

[SRD$^+$11] J.H. Song, M. Rensing, C.L.L.M. Daunt, P. O'Brien, and F.H. Peters. Directly modulated laser diode module exceeding 10 Gb/s transmission. *EEE Transactions on Components, Packaging and Manufacturing Technology*, 1(6):975 –980, Jun 2011.

[SRRI06] R.A. Shafik, S. Rahman, and AHM Razibul Islam. On the extended relationships among EVM, BER and SNR as performance metrics. In *International Conference on Electrical and Computer Engineering (ICECE)*, pages 408 –411, Dec 2006.

[SS86] Y. Silberberg and P. Smith. Subpicosecond pulses from a mode-locked semiconductor laser. *IEEE Journal of Quantum Electronics*, 22(6):759 – 761, Jun 1986.

[ST58] A. L. Schawlow and C. H. Townes. Infrared and optical masers. *Physical Review*, 112:1940–1949, Dec 1958.

[ST07] B.E.A. Saleh and M.C. Teich. *Fundamentals of Photonics*. Wiley-Interscience, 2007.

[STK96] M. Schell, M. Tsuchiya, and T. Kamiya. Chirp and stability of mode-locked semiconductor lasers. *IEEE Journal of Quantum Electronics*, 32(7):1180 –1190, Jul 1996.

[Sto98] J. Stott. The effects of phase noise in COFDM. `http://www.bbc.co.uk/rd/pubs/papers/pdffiles/jsebu276.pdf`, Summer 1998. Online; accessed 19-Sep-2012.

[Tec11] Agilent Technologies. Product Note 89400-14: Using error vector magnitude measurements to analyze and troubleshoot vector-modulated signals. http://cp.literature.agilent.com/litweb/pdf/5965-2898E.pdf , 2011. Online; accessed 19-Sep-2012.

[TEC12]   Ltd. Totoku Electric Co. Tcf series. Technical report, www.shf.de, 2012.

[TSG$^+$04]   Y. Takushima, H. Sotobayashi, M.E. Grein, E.P. Ippen, and H.A. Haus. Linewidth of mode combs of passively and actively mode-locked semiconductor laser diodes. *Proceedings of SPIE*, 5595:213–227, 2004.

[vD12]   F. van Dijk. Personal communication. Janurary 2012.

[vDEB$^+$08] F. van Dijk, A. Enard, X. Buet, F. Lelarge, and G. H. Duan. Phase noise reduction of a quantum dash mode-locked laser in a millimeter-wave coupled opto-electronic oscillator. *IEEE Journal of Lightwave Technology*, 26(15):2789–2794, Aug 2008.

[VHKKG97]   C.H. Von Helmolt, U. Kruger, K. Kruger, and G. Grosskopf. A mobile broad-band communication system based on mode-locked lasers. *IEEE Transactions on Microwave Theory and Techniques*, 45(8):1424–1430, Aug 1997.

[WED96]   K.J. Williams, R.D. Esman, and M. Dagenais. Nonlinearities in p-i-n microwave photodetectors. *IEEE Journal of Lightwave Technology*, 14(1):84 –96, Jan 1996.

[Wir10]   WirelessHD. Specification version 1.1 overview. `http://www.wirelesshd.org/pdfs/WirelessHD-Specification-Overview-v1.1May2010.pdf`, May 2010. Online; accessed 18-Sep-2012.

[WKYY$^+$00] E.L. Wooten, K.M. Kissa, A. Yi-Yan, E.J. Murphy, D.A. Lafaw, P.F. Hallemeier, D. Maack, D.V. Attanasio, D.J. Fritz, G.J. McBrien, and D.E. Bossi. A review of lithium niobate modulators for fiber-optic communications systems. *Selected Topics in Quantum Electronics, IEEE Journal of*, 6(1):69 –82, jan.-feb. 2000.

[WSPH91]   D. Wake, T.P. Spooner, S.D. Perrin, and I.D. Henning. 50 GHz InGaAs edge-coupled pin photodetector. *IET Electronics Letters*, 27(12):1073 –1075, Jun 1991.

[Yon11]   S.K Yong. *60 GHz Channel Characterizations and Modeling*, chapter 2, pages 17–61. John Wiley & Sons, 2011.

[YXVG11]   S.K. Yong, P. Xia, and Valdes-Garcia, editors. *60 GHz Technology for Gbps WLAN and WPAN: From Theory to Practice*, chapter 1, pages 1–16. John Wiley & Sons, 2011.

[Zap04]   H. Zappe. *Laser Diode Microsystems*. Springer Berlin, Heidelberg, New York, 1st edition, 2004.

[ZBCP]   T. Zwick, F. Brendel, B. Cabon, and J. Poette. European Patent Application No. 11 290 482.6: Radio communication system, home gateway, bidirectional communication system, and method for stabilising a sideband signal. Filed on Jan 19th, 2012.

[ZCY+11]   D. Zibar, A. Caballero, X. Yu, X. Pang, A.K. Dogadaev, and I.T. Monroy. Hybrid optical fibre-wireless links at the 7-110 GHz band supporting 100 Gbps transmission capacities. In *International Topical Meeting on Microwave Photonics (MWP)/ Asia-Pacific Microwave Photonics Conference (APMP), Singapore*, pages 445 – 449, Oct 2011.

For Saffar Arjmandi